Habib Ammari, Bryn Davies

Metamaterial Analysis and Design

De Gruyter Series in Applied and Numerical Mathematics

Volume 9

Habib Ammari, Bryn Davies

Metamaterial Analysis and Design

A Mathematical Treatment of Cochlea-inspired Sensors

DE GRUYTER

Mathematics Subject Classification 2020
Primary: 35J05, 35C20, 35P20, 94A12; Secondary: 35R30, 74J20, 78A45, 35Q92, 37N25, 92C20

Authors
Prof. Habib Ammari
ETH Zürich
Department of Mathematics
Rämistrasse 101
8092 Zürich
Switzerland
habib.ammari@math.ethz.ch

Dr. Bryn Davies
Imperial College London
Department of Mathematics
Huxley Building
South Kensington Campus
London SW7 2AZ
UK
bryn.davies@imperial.ac.uk

ISBN 978-3-11-078404-6
e-ISBN (PDF) 978-3-11-078496-1
e-ISBN (EPUB) 978-3-11-078513-5
ISSN 2512-1820

Library of Congress Control Number: 2023942988

Bibliographic information published by the Deutsche Nationalbibliothek
The Deutsche Nationalbibliothek lists this publication in the Deutsche Nationalbibliografie; detailed bibliographic data are available on the Internet at http://dnb.dnb.de.

Typesetting: VTeX UAB, Lithuania
Printing and binding: CPI books GmbH, Leck

www.degruyter.com

Contents

1 Introduction

The emergence of metamaterials has allowed scientists and engineers to interact with and control waves in previously unforeseen ways. By combining two or more materials in elaborate composite structures, it is possible to reach new, previously inaccessible regions of material property space [29]. The addition of several geometric parameters, as well as the material parameters of the constituent elements, gives engineers newfound freedom to fine-tune the properties of their materials.

The term "metamaterial" was introduced in 2000 by Rodger Walser [165] as a catch-all term for heterogeneous materials whose properties go far beyond those of their constituent materials. In particular, metamaterials are those that achieve an "optimized combination, not available in nature, of two or more responses to specific excitation." This somewhat science-fiction notion of metamaterials displaying out-of-this-world properties is indicative of the discoveries that initiated the field, which included flat lenses [137, 160] and invisibility cloaks [124, 55]. While this space-age characterization of metamaterials persists to this day (at the time of writing, the first sentence of the "Metamaterial" Wikipedia page introduces them as "any material engineered to have a property that is not found in naturally occurring materials" [167]), metamaterials are increasingly becoming everyday objects.

As the field of metamaterials has matured, in the two decades since its inception, notions of what it means to be a metamaterial have varied. While a universal definition is yet to emerge, a key idea is that they are composed of constituent units that are smaller than the operating wavelength. In general, wave equations with rapidly oscillating coefficients are often studied using homogenization theory, whereby asymptotic methods are used to derive the *effective* properties of the macroscopic material within certain regimes [39, 98]. This is a powerful approach, that underpins much of modern metamaterial research, for which the canonical problem is to control the effective properties of a metamaterial by manipulating the geometry of the subwavelength units.

The main objects of study in this book will be high-contrast metamaterials whose constituent units experience *subwavelength resonance*. That is, they experience resonance in response to wavelengths that are much greater than their size. This phenomenon is a consequence of the high-contrast regime considered here and is reminiscent of the *Minnaert resonance* of an air bubble in water [127]. These resonators behave like a multidimensional spring, with the pressure inside the cavity yielding a restorative force in response to pressure waves in the surrounding medium, that causes them to oscillate in a so-called breathing mode [67, 80]. Crucially, it is the large material contrast that leads to the deep subwavelength response. In the case of an air bubble in water, air is around 1000 times less dense than water and the Minnaert resonance occurs at a wavelength that is approximately 100 times the bubble's radius. Acoustic metamaterials based on this principle can be fabricated, for example, by injecting air bubbles into polymer gels [108, 109]. Similar deeply subwavelength responses have also

https://doi.org/10.1515/9783110784961-001

been observed in other high-contrast systems, such as dielectric nanoparticles [19, 107], plasmonic nanoparticles [20], and Helmholtz resonators [21].

The presence of local (subwavelength) resonances in a metamaterial presents an obstacle for applying traditional homogenization theories. Even though the constituent units are smaller than the operating wavelength, their effect on wave propagation is large because of the local resonances. Hence, standard homogenization techniques are not directly applicable and the derivation of effective properties of metamaterials becomes especially challenging in the limit that the contrast of the ingredient materials diverges as the cell size shrinks to zero [98]. Recently, homogenization techniques have been introduced for deriving effective properties of infinitely periodic systems of locally resonant unit cells, including for the specific high-contrast case of Minnaert resonances [77, 99, 22]. In this book, we will similarly take advantage of the highly contrasting parameters to perform asymptotic analyses of the wave scattering problem. In our case, we will consider a high-contrast, low-frequency asymptotic limit and derive convenient characterizations of the resonant modes of the system in terms of the eigenstates of the *generalized capacitance matrix* [7].

In this book, we will explore how asymptotic techniques can be used to analyze and design novel metamaterials efficiently, and without the need for enormous computational expense. As a case study, to center this work, we will study a collection of problems related to hearing and cochlear function. The cochlea's position inside the human head and the key features of its role in hearing are shown in Figure 1.1. It plays a crucial role as it is here that mechanical waves are converted into neural signals. Additionally, the cochlea also contributes to auditory processing by filtering waves according to their frequency. If this function can be replicated with metamaterials, it would open novel possibilities for scientists and engineers.

The cochlea's spatial frequency separation is due to its graded dimensions and material parameters. In simple terms, the cochlea is a long, fluid-filled tube that has an elastic membrane suspended down its center. This membrane has around 3500 receptor cells mounted upon it. Both the width and the stiffness of this membrane are graded along the length of the cochlea, such that it is narrow and stiff at the base and wider and more mobile at the apex. This grading means that the membrane interacts differently with different frequencies at different positions. As a result, high frequencies give a peak response at one end while lower frequencies lead to a peak response at the other end of the membrane. This means the body can identify the frequency of a pure tone from the position of the receptor cells that give the maximal response.

Frequency separation in graded structures is not unique to the cochlea and there is a sizeable body of research devoted to developing graded metamaterials that perform similar functions. By varying either the dimensions or the spacing of the metamaterial's constituent units, it is similarly possible to vary how it interacts with different frequencies along the structure. The most widely utilized consequence of this is the phenomenon of *rainbow trapping* [158]. If the metamaterial is one that has a spectral gap (a range of frequencies that cannot propagate through the structure), then this gap can be shifted up

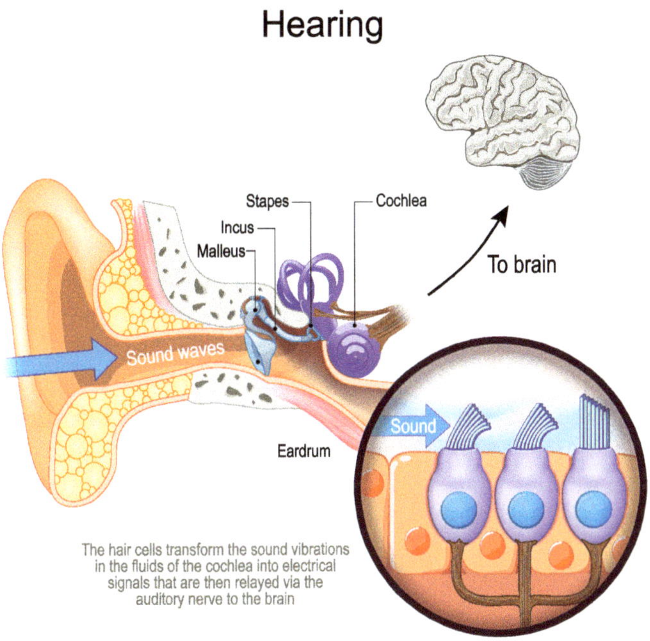

Figure 1.1: The cochlea is the central organ of human hearing. It has a distinctive spiral shape and graded parameters, allowing it to spatially filter different sounds according to their frequency. The filtered sounds cause deflections in the stereocilia of receptor cells. These mechanical deflections are subsequently converted to neural signals that are transmitted to the brain. Picture credit: ttsz / iStock / Getty Images.

or down in frequency due to the gradient. This means different frequencies are stopped from propagating (and, hence, "trapped") at different positions, giving a "rainbow" of different frequencies. After its initial realization in optics [158], it has since been implemented in settings ranging from acoustics [176] and elasticity [26, 148] to seismic waves [52] and water waves [38]. Having applications in machine hearing, as we will explore in detail in this book, graded metamaterials have also been used to great effect in building energy harvesting devices [65], where the rainbow trapping effect is used to focus wave energy so that it can be harvested more efficiently [66, 174].

Given the qualitative similarities between the functions performed by graded metamaterials and the cochlea, a natural question to ask is whether this correspondence can be made more exact. If so, can we leverage this connection to draw useful scientific conclusions? In order to answer these questions, we need to develop concise, rigorous methods for characterizing the graded metamaterial's properties.

Designing graded metamaterials that mimic the function of the cochlea is an example of *biomimicry* [40]. This is the scientific practice of designing systems to replicate the function of biological systems. This seeks to take advantage of the fact that, thanks to millennia of evolution, natural systems often present innovative and finely-tuned solutions to challenging problems. The principle of biomimicry has been used in many different settings, including the aerodynamic design of high-speed trains, the creation of passive cooling systems in buildings and the invention of hook-and-loop fasteners such as Velcro [92]. This methodology has also found its way into metamaterial science, with metamaterials having been inspired by many different systems, such as moth wings [131], spider webs [126, 57], and bone [81, 125], as well as the cochlea.

Many different examples of cochlea-inspired devices and systems have been developed. These range from scaled-up versions of individual cochlear receptor cells [94] to table-top realizations of active models [146] and elaborate signal processing architectures [116]. These all feature the crucial component of spatially graded constituent elements, which are typically resonant. There are different choices for the resonant elements: Helmholtz resonators [101], quarter-wavelength resonators [146], and Minnaert resonators [4] have all been used. The elements are typically arranged in a straight line, but can also be coiled into a spiral to replicate the compact shape of the cochlea [175]. Useful insight has also been gained from considering abstracted models, such as those composed of coupled masses and springs [32] or vibrating reeds [36]. While most of the systems are passive and do not have any energy inputs, the cochlea uses a nonlinear amplification mechanism and active metamaterials have been developed to replicate this [146, 5, 61].

Another important connection to explore is the link between cochlea-inspired sensors and signal processing algorithms. Here, we have two (sometimes disconnected) communities both aiming to achieve the same feat: designing and realizing systems that replicate how humans and other animals hear. Realizing this link has facilitated breakthroughs in both communities, through a perfect example of biomimicry in action [116]. This idea can be applied to cochlea-inspired metamaterials. If we have an analytic formula that describes how the metamaterial behaves in response to a given incoming acoustic signal, then this formula can already be viewed as a simple signal processing algorithm [6]. This simple algorithm can then be developed further by taking ideas either from established signal processing practices or additional bio-inspired concepts. Putting this all together, we have realized a three-way exchange of ideas between biological auditory systems, acoustic metamaterials and signal processing algorithms, as depicted in Figure 1.2.

This book will begin by outlining the key concepts that underpin graded metamaterials. This can be found in Chapter 2, where we summarize the key ideas needed to understand the spectra of both perfectly periodic structures and those that have been subjected to a monotonic gradient function. We will then move on to studying the high-contrast graded metamaterials that are the main object of study in this book. We will consider passive two- and three-dimensional structures in Chapters 3 and 4, respectively. In each case, we will develop the asymptotic and numerical methods that are needed in order to be able to design the cochlea-inspired structures we are looking for. In Chapter 5 we will consider the implications of cochlea-inspired metamaterials for bio-inspired signal processing. We will use the asymptotic formulas which characterize cochlea-inspired metamaterials as the first step of a signal processing algorithm, and

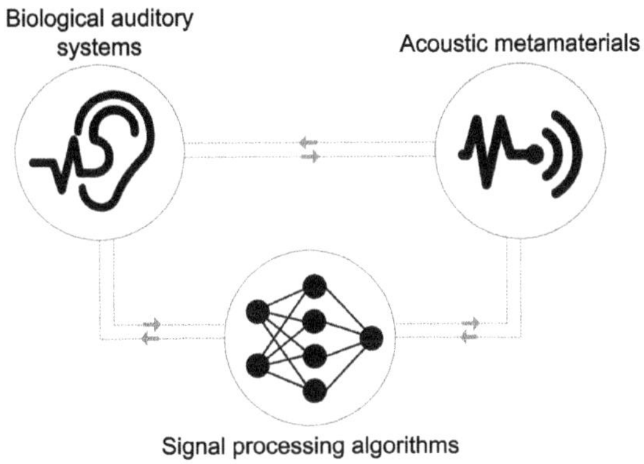

Figure 1.2: The design of hearing-inspired acoustic metamaterials and signal processing algorithms is an example of biomimicry. Given enough data or analytic formulas, we can explore the three-way exchange of design principles and features between these three different types of systems.

propose some additional bio-inspired features. In Chapter 6 we will study the robustness of the subwavelegth cochlea-like metamaterials considered here with respect to random imperfections and defects. This is important for the viability of physical devices, which are inevitably going to have small defects introduced either by manufacturing processes or through intensive usage. Finally, in Chapter 7, we will briefly consider active cochlea-inspired metamaterials, to examine if adding nonlinear amplification mechanisms to the system can better replicate the function of the cochlea. Chapter 8 contains some concluding remarks and discusses some of the priorities for future investigation in the field of cochlea-inspired graded metamaterials.

2 Graded metamaterials

2.1 Spectral band diagrams

Before considering graded structures, we first consider periodic arrangements of units. In this case, there are a suite of techniques known as *Floquet–Bloch* analysis that can be used to describe the spectrum very efficiently. An excellent review of these techniques for periodic elliptic operators can be found in [106]. The key idea behind this approach is that instead of studying the periodic problem, we can apply the *Floquet transform* to study a collection of *α-quasiperiodic* problems whose spectra are equivalent to the original problem but are easier to understand.

Suppose that we study a d-dimensional differential problem and have a material with a d_l-dimensional periodic lattice Λ. Let $l_1, \dots, l_{d_l} \in \mathbb{R}^d$ be the lattice vectors that generate the lattice Λ, i. e.,

$$\Lambda := \{m_1 l_1 + \cdots + m_{d_l} l_{d_l} \mid m_i \in \mathbb{Z}\}.$$

For simplicity, we assume that the lattice is aligned with the first d_l coordinate axes, so that a point $x \in \mathbb{R}^d$ can be written using the notation $x = (x_l, x_0)$, where $x_l \in \mathbb{R}^{d_l}$ is the vector along the first d_l dimensions and $x_0 \in \mathbb{R}^{d-d_l}$. Denote by $Y \subset \mathbb{R}^d$ a fundamental domain of the given lattice:

$$Y := \{c_1 l_1 + \cdots + c_{d_l} l_{d_l} \mid 0 \leq c_1, \dots, c_{d_l} \leq 1\}.$$

Two examples of lattices and unit cells are sketched in Figure 2.1, a line array with $d_l = 1$ and square array with $d_l = 2$.

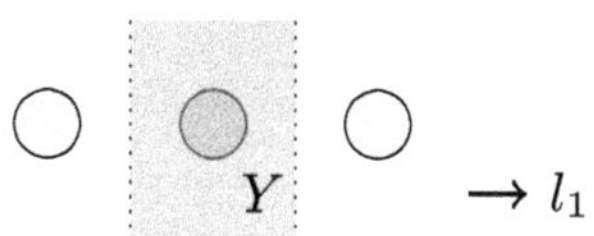

(a) A line array

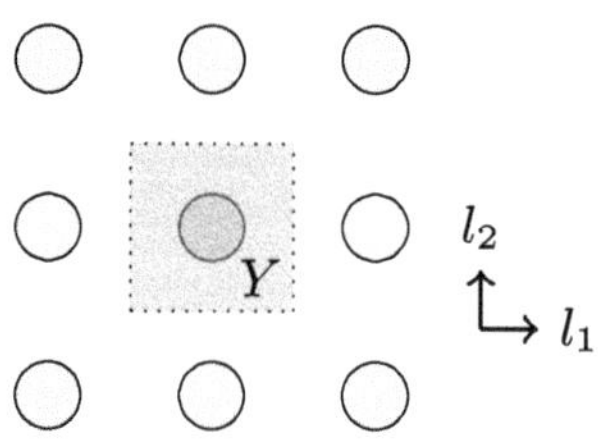

(b) A square array

Figure 2.1: Two examples of periodic lattices.

A key concept for understanding how periodic materials interact with waves is their *dual* or *reciprocal lattice*. The dual lattice of Λ, denoted by Λ^*, can be thought of as the Fourier transform of Λ and is generated by the vectors $\alpha_1, \dots, \alpha_{d_l} \in \mathbb{R}^d$ satisfying $\alpha_i \cdot l_j = 2\pi\delta_{ij}$, where δ_{ij} is the Kronecker delta. This condition determines the first d_l coordinates of α_i, and we impose that the last $d - d_l$ coordinates are zero, for uniqueness.

https://doi.org/10.1515/9783110784961-002

To see the relevance of the reciprocal lattice, we must introduce the Floquet transform. We first make an important definition, of what it means for a function to be *quasiperiodic.*

Definition 2.1.1. Given $\alpha \in \mathbb{R}^d$, a function $f : \mathbb{R}^d \to \mathbb{C}$ is said to be α-quasiperiodic if $e^{-\mathrm{i}\alpha\cdot x} f(x)$ is periodic.

Now, given a function $f \in L^2(\mathbb{R}^d)$ and a lattice $\Lambda \subset \mathbb{R}^d$, the Floquet transform of f is defined as

$$\mathcal{F}[f](x,\alpha) := \sum_{m\in\Lambda} f(x-m)e^{\mathrm{i}\alpha\cdot m}, \quad x,\alpha \in \mathbb{R}^d. \tag{2.1}$$

By construction, $\mathcal{F}[f]$ is always α-quasiperiodic in x and periodic in α. For more details on the Floquet transform, see, for example, [14, 105].

If $\mathcal{F}[f](x,\alpha)$ is α-quasiperiodic in x along the directions of the lattice Λ, then the periodicity in α will be along the directions of the reciprocal lattice Λ^*. The unit cell of this periodicity in reciprocal space is known as the *Brillouin zone*. The Brillouin zone Y^* is defined as $Y^* := (\mathbb{R}^{d_l} \times \{\mathbf{0}\})/\Lambda^*$, where $\mathbf{0}$ is the zero-vector in $\mathbb{R}^{d-d_l}$. We remark that Y^* can be written as $Y^* = Y_l^* \times \{\mathbf{0}\}$, where Y_l^* has the topology of a d_l-dimensional torus.

The premise of Floquet–Bloch analysis is that the spectrum of a periodic elliptic differential operator can be decomposed by applying the Floquet transform. As an example, let $D \subset \mathbb{R}^d$ be a bounded, connected set with appropriately smooth boundary and

$$\mathcal{D} = \bigcup_{m\in\Lambda} D + m, \tag{2.2}$$

be the periodic structure consisting of D repeated infinitely many times. Then, suppose we wish to solve the problem

$$\begin{cases} \Delta u + \frac{\omega^2}{v^2} u = 0 & \text{in } \mathbb{R}^d \setminus \mathcal{D}, \\ \Delta u + \frac{\omega^2}{v_b^2} u = 0 & \text{in } \mathcal{D},\ i = 1,\dots,N, \\ u|_+ - u|_- = 0 & \text{on } \partial\mathcal{D}, \\ \delta \frac{\partial u}{\partial \nu}\big|_+ - \frac{\partial u}{\partial \nu}\big|_- = 0 & \text{on } \partial\mathcal{D},\ i = 1,\dots,N, \\ u(x_l, x_0) & \text{satisfies a radiation condition as } |x_0| \to \infty, \end{cases} \tag{2.3}$$

where v and v_b are real wave speeds and δ is some real-valued material contrast parameter. The appropriate radiation condition is a condition on the solution in the far field which guarantees that energy can only propagate outwards from the structure, and is sufficient for the problem to be well posed. The *spectrum* of (2.3) is the set σ of all $\omega \in \mathbb{C}$ such that (2.3) has a nonzero solution u. We can characterize the spectrum by applying the Floquet transform to (2.3), yielding a differential problem in terms of $u^\alpha(x) := \mathcal{F}[u](x,\alpha)$, given by

$$\begin{cases} \Delta u^\alpha + \frac{(\omega^\alpha)^2}{v^2} u^\alpha = 0 & \text{in } \mathbb{R}^d \setminus \mathcal{D}, \\ \Delta u^\alpha + \frac{(\omega^\alpha)^2}{v_b^2} u^\alpha = 0 & \text{in } \mathcal{D},\ i = 1, \dots, N, \\ u^\alpha|_+ - u^\alpha|_- = 0 & \text{on } \partial\mathcal{D}, \\ \delta \frac{\partial u^\alpha}{\partial \nu}\big|_+ - \frac{\partial u^\alpha}{\partial \nu}\big|_- = 0 & \text{on } \partial\mathcal{D},\ i = 1, \dots, N, \\ u^\alpha(x_d, x_0) & \text{is } \alpha\text{-quasiperiodic in } x_d, \\ u^\alpha(x_d, x_0) & \text{satisfies a radiation condition as } |x_0| \to \infty. \end{cases} \tag{2.4}$$

This new differential problem is much more convenient to study. The set of all ω_k^α for which there exists a non-trivial solution to (2.4) is known as the *Bloch spectrum* of (2.4), which we denote as $\sigma(\alpha)$. It is generally the case that ω_k^α is a Lipschitz continuous function of α, meaning that each eigenvalue ω_k^α traces out a curve as α is varied. These curves are often referred to as (spectral) *bands*.

Given the Bloch spectrum $\{\omega_k^\alpha : k = 1, 2, \dots, \alpha \in Y^*\}$ of (2.4), we would like to recover the spectrum of the original operator (2.3). Fortunately, the Floquet transform is an invertible map $\mathcal{F} : L^2(\mathbb{R}^d) \to L^2(Y \times Y^*)$, with inverse given by

$$\mathcal{F}^{-1}[g](x) = \frac{1}{|Y_l^*|} \int_{Y^*} g(x, \alpha)\, \mathrm{d}\alpha, \quad x \in \mathbb{R}^d, \tag{2.5}$$

where $g(x, \alpha)$ is α-quasiperiodic in x. As result, we are able to recover solutions of (2.3). In particular, it turns out we can recover the spectrum of (2.3) by taking the union of the Bloch spectra of (2.4) over all the possible quasiperiodicities α. That is,

$$\sigma = \bigcup_{\alpha \in Y^*} \sigma(\alpha), \quad \text{where } \sigma(\alpha) = \bigcup_{i=1}^{\infty} \omega_i^\alpha. \tag{2.6}$$

Knowing the spectrum σ of a material will be immensely useful when it comes to building waveguides and other wave control devices. In particular, if a given frequency ω is not part of the spectrum σ, then it is not able to propagate through the material. Instead, a wave of frequency ω will have an exponentially decaying amplitude within the material as its energy is reflected. The ranges of frequencies for which this occurs are known as *band gaps*, as they are the gaps between the continuous spectral bands. We make this precise with the following definition.

Definition 2.1.2 (Band gap). In general, a band gap of the periodic structure $\mathcal{D}$ is a connected component of $\mathbb{C} \setminus \sigma$. If the spectrum σ is real, then we define a band gap of $\mathcal{D}$ as a connected component of $\mathbb{R} \setminus \sigma$, in which case it consists of a set of intervals.

It is helpful to demonstrate this approach with an example. Suppose that $d = 3$ and D is a sphere. Consider a one-dimensional periodic array ($d_l = 1$) with period L. In this case, the Brillouin zone is $Y_l^* \simeq [-\pi/L, \pi/L)$ and has the topology of a circle. The real parts of the bands for a simple case of a single spherical inclusion is shown in

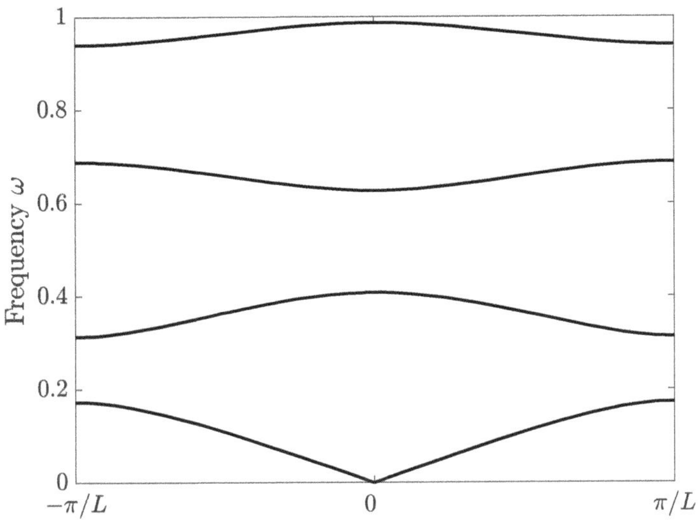

Figure 2.2: The real parts of the spectral band structure of a one-dimensional array. An array of spherical resonators is modeled, as sketched in Figure 2.1a, which has a series of spectral bands, with band gaps between them.

Figure 2.2. Features to notice are that the lowest band has $\omega_1^\alpha \to 0$ as $\alpha \to 0$, and that it does so approximately linearly. Additionally, each band has zero slope at the edges of the Brillouin zone, where $\alpha = \pm\pi/L$. This is a standard phenomenon and is due to the so-called *Bragg condition* [43]. This simple system has been chosen to exhibit a series of band gaps between each of the first few spectral bands.

The analysis is similar for a two-dimensional periodic array ($d_l = 2$). In this case, the Brillouin zone $Y_l^* \simeq [-\pi/L, \pi/L) \times [-\pi/L, \pi/L)$ and has the topology of a (two-dimensional) torus. The Brillouin zone is sketched on the right in Figure 2.3. Thanks to the symmetries of the spherical inclusions, it is sufficient to study just part of the Brillouin zone. This part is known as the *irreducible Brillouin zone* and is the shaded triangle in Figure 2.3. For any other point within the square $[-\pi/L, \pi/L) \times [-\pi/L, \pi/L)$, the corresponding differential problem is equivalent to that given by a point in the irreducible Brillouin zone. For most standard examples, it will be the case that the maxima and minima of each band will occur at the edges of the irreducible Brillouin zone, on the curve ΓMXΓ. While this approach loses some important information about the dispersive properties of the periodic structure [53], it is a widespread convention in the literature and means that the multidimensional band functions (which are surfaces) can be plotted as one-dimensional curves, as we do in Figure 2.3. This configuration has a band gap between the first and second bands (between approximately $\omega = 0.95$ and $\omega = 1.3$ in Figure 2.3).

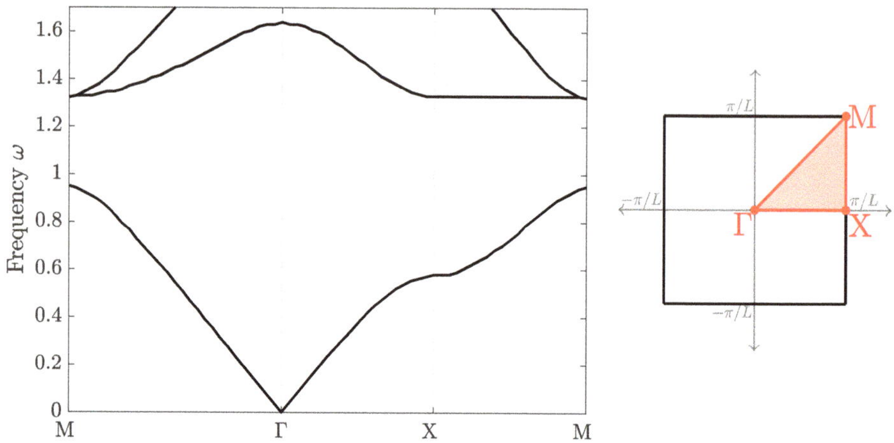

Figure 2.3: The real parts of the spectral band structure of a two-dimensional square array. An array of spherical resonators is modeled, as sketched in Figure 2.1b. We plot a dispersion curve around the edges of the irreducible Brillouin zone, as sketched on the right. There is a band gap between the first and second bands.

2.2 Graded parameters

One way to achieve interesting wave-control effects with periodic structures is to vary the structure's material parameters. In particular, it is possible to create materials that experience different effective band gaps in different spatial regions. This is the main idea behind *graded* metamaterials: to vary the dimensions or parameters of the constituent elements in order to shift the local band gap along the array. Provided that this variation happens relatively slowly, so as not to introduce any impedance mismatches, then the metamaterial's local properties can be understood by looking at the equivalent periodic material. This formalism can be made precise using asymptotic techniques such as high frequency homogenization, as developed in [56] and applied to graded problems, e. g., by [147, 25, 63, 54].

The most common type of graded metamaterial is based on a monotonic gradient function. This has the effect of slowly shifting the effective band gap as a function of position in the array. An example is shown in Figure 2.4. In this case, we have gradually increased the radius of the scatterers along the array. The plot shows the position of the first band gap as a function of the diameter for the same one-dimensional array that was considered in Figure 2.2 and sketched in Figure 2.1a. The band gap shifting downwards with increasing diameter means that if a wave is incident on the array from the left, then high frequencies will experience an effective band gap before lower frequencies do. This establishes a monotonic relationship between the frequency of a pure tune and the position at which it is reflected (due to the effective band gap). This is the rainbow trapping effect in action.

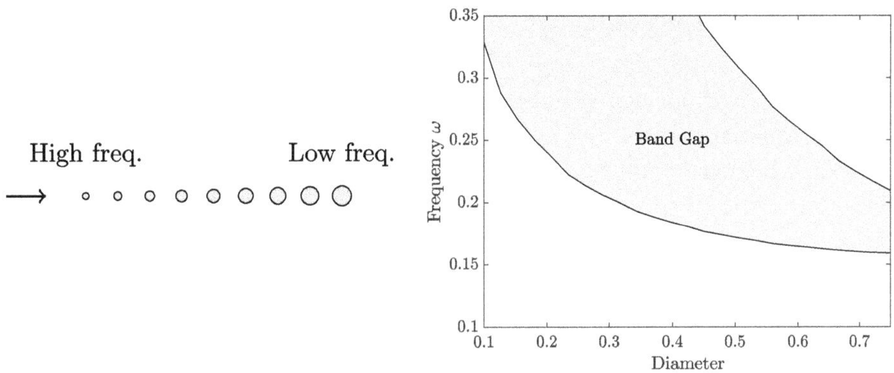

Figure 2.4: Graded metamaterials produce a rainbow effect by gradually shifting the effective band gap as a wave propagates along the array, meaning different frequencies are cut off (and reflected) at different positions. Here, we plot the first band gap for a one-dimensional array of spheres for different diameters.

Gradients that have turning points are also interesting and will lead to wave localization. For example, symmetric gradient functions can be interpreted as placing two monotonically graded metamaterials back to back, for example, which would mean that waves of specified frequencies are unable to propagate in either direction and would be localized to the center [25, 63]. This phenomenon is important as it can be used as the starting point for building waveguides, however, we will not explore it in detail here.

3 Metasurface design

We will begin by studying two-dimensional models. While the asymptotic analysis is slightly more challenging in two dimensions than in three, the problem is much easier to handle numerically, making this an ideal starting point.

3.1 Problem formulation

We consider a domain D in $\mathbb{R}^2$ which is the disjoint union of $N \in \mathbb{N}$ bounded and simply connected subdomains $\{D_1, \ldots, D_N\}$. We suppose that each boundary ∂D_n is such that there is some $0 < s < 1$ so that $\partial D_n \in C^{1,s}$ (that is, each ∂D_n is locally the graph of a differentiable function whose derivatives are Hölder continuous with exponent s). We will consider the resonators arranged in a straight line. Figure 3.1 shows an example of such an arrangement.

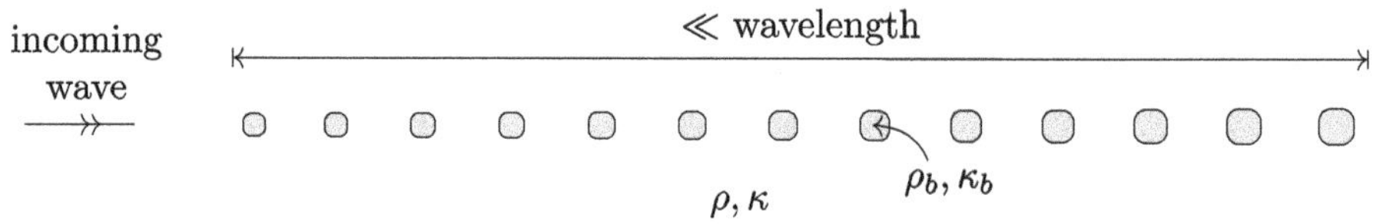

Figure 3.1: A graded array of subwavelength resonators. The large density contrast $\delta := \rho_b/\rho \ll 1$ is crucial for the subwavelength resonant response of the structure.

We denote by ρ_b and κ_b the density and bulk modulus of the interior of the material inclusions, respectively. Meanwhile, ρ and κ are used for the corresponding parameters for the surrounding fluid, which we assume occupies $\mathbb{R}^2 \setminus \overline{D}$. These quantities are indicated on Figure 3.1. We can then introduce the auxiliary parameters

$$v = \sqrt{\frac{\kappa}{\rho}}, \quad v_b = \sqrt{\frac{\kappa_b}{\rho_b}}, \quad k = \frac{\omega}{v}, \quad k_b = \frac{\omega}{v_b}, \tag{3.1}$$

which are the wave speeds and wavenumbers in $\mathbb{R}^2 \setminus \overline{D}$ and in D, respectively. Finally, we also introduce the dimensionless contrast parameter δ as

$$\delta = \frac{\rho_b}{\rho}. \tag{3.2}$$

The appearance of a small δ in the equations is the cause of the system's subwavelength resonant response and will be at the center of our subsequent asymptotic analysis.

https://doi.org/10.1515/9783110784961-003

We will consider the scattering of a time-harmonic acoustic wave by this array of material inclusions. This is described by the Helmholtz system of equations

$$\begin{cases} (\Delta + k^2)u(x,\omega) = 0 & \text{in } \mathbb{R}^2 \setminus \overline{D}, \\ (\Delta + k_b^2)u(x,\omega) = 0 & \text{in } D, \\ u_+ - u_- = 0 & \text{on } \partial D, \\ \delta \frac{\partial u}{\partial \nu}\big|_+ - \frac{\partial u}{\partial \nu}\big|_- = 0 & \text{on } \partial D, \\ u^s := u - u^{\text{in}} \text{ satisfies the SRC} & \text{as } |x| \to \infty, \end{cases} \tag{3.3}$$

where u^{in} is the incoming wave, $\frac{\partial}{\partial \nu}$ denotes the outward normal derivative and the subscripts + and − are used to denote evaluation from outside and inside ∂D respectively. “SRC” is used to denote the Sommerfeld radiation condition

$$\lim_{|x|\to\infty} |x|^{1/2}\left(\frac{\partial}{\partial |x|} - \mathrm{i}k\right)u^s(x,\omega) = 0. \tag{3.4}$$

The system of equations (3.3) can be interpreted as follows. We have a Helmholtz equation, which is an eigenvalue problem for the Laplacian Δ, in both the interior and exterior of the material inclusions. This is obtained by substituting a time-harmonic ansatz $e^{-\mathrm{i}\omega t}u(x)$ into the scalar wave equation, to separate the dependence on time and space. The conditions on ∂D give the continuity of the field and the continuity of flux across the boundary. Finally, the Sommerfeld radiation condition is the condition required to ensure that we select the solution that is outgoing (rather than incoming from infinity) and gives the well-posedness of the problem (3.3).

We will use integral operators known as layer potentials to represent the solutions to the scattering problem (3.3). This will allow us to describe a broad class of different shapes of material inclusions concisely. Layer potentials are Green’s function operators, for which we need to define the outgoing fundamental solution G^k on $\mathbb{R}^2$ as the unique solution to

$$(\Delta + k^2)G^k(x) = \delta_0(x), \tag{3.5}$$

where δ_0 is the Dirac delta function and G^k is assumed to satisfy the Sommerfeld radiation condition in the far field, as $|x| \to \infty$. In two dimensions, it can be shown that G^k is given by

$$G^k(x) = -\frac{\mathrm{i}}{4}H_0^{(1)}(k|x|), \tag{3.6}$$

where $H_0^{(1)}$ is the Hankel function of the first kind and order zero.

Definition 3.1.1. We define the Helmholtz single layer potential associated with the domain D and wavenumber k as

$$\mathcal{S}_D^k[\varphi](x) = \int_{\partial D} G^k(x-y)\varphi(y)\,\mathrm{d}\sigma(y), \quad x \in \partial D, \varphi \in L^2(\partial D). \tag{3.7}$$

We similarly define the Neumann–Poincaré operator associated with D and k as

$$\mathcal{K}_D^{k,*}[\varphi](x) = \int_{\partial D} \frac{\partial G^k(x-y)}{\partial \nu(x)}\varphi(y)\,\mathrm{d}\sigma(y), \quad x \in \partial D, \varphi \in L^2(\partial D). \tag{3.8}$$

Given the definitions of the single layer potential (3.7) and the Neumann–Poincaré operator (3.8), we can represent the solution to (3.3) as

$$u = \begin{cases} u^{\text{in}}(x) + \mathcal{S}_D^k[\psi](x), & x \in \mathbb{R}^2 \backslash \overline{D}, \\ \mathcal{S}_D^{k_b}[\phi](x), & x \in D, \end{cases} \tag{3.9}$$

for some surface potentials $(\phi, \psi) \in L^2(\partial D) \times L^2(\partial D)$ [18]. A consequence of this ansatz is that three of the five conditions in (3.3) are automatically satisfied. Thanks to the choice of Green's function in the definition of $\mathcal{S}_D^k$, the function u given by (3.9) necessarily satisfies the Helmholtz equations on the interior and exterior as well as the Sommerfeld radiation condition in the far field. Thus, it remains only to find $(\phi, \psi) \in L^2(\partial D) \times L^2(\partial D)$ such that the two transmission conditions on ∂D are satisfied. The continuity of flux condition in (3.3) can be restated using the well-known property that [18]

$$\frac{\partial}{\partial \nu}\mathcal{S}_D^k[\varphi]|_{\pm} = \left(\pm\frac{1}{2}I + \mathcal{K}_D^{k,*}\right), \tag{3.10}$$

where I denotes the identity on $L^2(\partial D)$. Therefore, the problem (3.3) is equivalent to finding $(\phi, \psi) \in L^2(\partial D) \times L^2(\partial D)$ such that

$$\mathcal{A}(\omega, \delta)\begin{pmatrix}\phi \\ \psi\end{pmatrix} = \begin{pmatrix} u^{\text{in}} \\ \delta\frac{\partial u^{\text{in}}}{\partial \nu}\end{pmatrix}, \tag{3.11}$$

where

$$\mathcal{A}(\omega, \delta) := \begin{bmatrix} \mathcal{S}_D^{k_b} & -\mathcal{S}_D^k \\ -\frac{1}{2}I + \mathcal{K}_D^{k_b,*} & -\delta(\frac{1}{2}I + \mathcal{K}_D^{k,*}) \end{bmatrix}. \tag{3.12}$$

Here, $\mathcal{A}(\omega, \delta)$ is a map from $L^2(\partial D) \times L^2(\partial D)$ to $H^1(\partial D) \times L^2(\partial D)$, with $L^2(\partial D)$ being the standard space of functions that are square integrable on the boundary ∂D, while $H^1(\partial D)$ is the subset of $L^2(\partial D)$ which contains functions that have weak first derivatives that are also square integrable.

In order to understand the behavior of the system of acoustic resonators, we will primarily be interested in characterizing the resonant modes of the system.

Definition 3.1.2. A resonant frequency is defined to be $\omega \in \mathbb{C}$ such that there exists a nontrivial solution to

$$\mathcal{A}(\omega,\delta)\begin{pmatrix}\phi\\ \psi\end{pmatrix} = \begin{pmatrix}0\\ 0\end{pmatrix}, \tag{3.13}$$

where $\mathcal{A}(\omega,\delta)$ is defined in (3.12). For each resonant frequency ω, we define the corresponding eigenmode (or resonant mode) as

$$u = \begin{cases} \mathcal{S}_D^k[\psi](x), & x \in \mathbb{R}^2 \setminus D, \\ \mathcal{S}_D^{k_b}[\phi](x), & x \in D. \end{cases} \tag{3.14}$$

3.2 Asymptotic analysis

Before we start considering complex metamaterials consisting of interesting arrangements of our resonators, we wish to develop an efficient analytic method for characterizing their behavior. Our approach will be to perform asymptotic analysis in terms of the material contrast parameter δ.

We are interested in a regime where the contrast of material densities is large but the wave speeds are of the same order. That is, if we have the two dimensionless contrast parameters

$$\delta = \frac{\rho_b}{\rho} \quad \text{and} \quad \tau = \frac{k_b}{k} = \frac{v_b}{v} = \sqrt{\frac{\rho\kappa_b}{\rho_b\kappa}}, \tag{3.15}$$

then we assume that

$$v = O(1), \quad v_b = O(1) \quad \text{and} \quad \tau = O(1). \tag{3.16}$$

We also assume that the rescaled dimensions are such that the subdomains $\{D_1,\dots,D_N\}$ have widths that are $O(1)$. On the other hand, we assume that there is a large contrast between the densities inside and outside the resonators, so that

$$\delta \ll 1. \tag{3.17}$$

In addition to assuming that the material contrast is large (in the sense that $\delta \to 0$) we will consider a low-frequency limit. This is motivated by the fact that we are interested in metamaterial structures in which the operating wavelengths are expected to be much larger than the dimensions of the locally resonant elements. Further, we wish to exploit the interesting subwavelength resonance (Minnaert resonance) that occurs in this regime. With this in mind, we make the following definition of a *subwavelength* resonant frequency:

Definition 3.2.1. A resonant frequency $\omega = \omega(\delta)$ is said to be a subwavelength resonant frequency if it depends continuously on δ and satisfies

$$\omega(\delta) \to 0 \quad \text{as} \quad \delta \to 0.$$

The regime in which the frequency is asymptotically low is helpful for our analysis, as it means that we can take advantage of known asymptotic expansions of the operator $\mathcal{A}$, given in (3.12).

Lemma 3.2.1. *In the space of bounded linear operators from $L^2(\partial D) \times L^2(\partial D)$ to $H^1(\partial D) \times L^2(\partial D)$, we have that*

$$\mathcal{A}(\omega, \delta) = \mathcal{A}_0 + \omega^2 \ln \omega \mathcal{A}_{1,1,0} + \omega^2 \mathcal{A}_{1,2,0} + \delta \mathcal{A}_{0,1} + O(\delta\omega^2 \ln \omega) + O(\omega^4 \ln \omega),$$

as $\omega, \delta \to 0$, where

$$\mathcal{A}_0 := \begin{bmatrix} \hat{\mathcal{S}}_D^{k_b} & -\hat{\mathcal{S}}_D^{k} \\ -\frac{1}{2}I + \mathcal{K}_D^* & 0 \end{bmatrix}, \quad \mathcal{A}_{1,1,0} := \begin{bmatrix} v_b^{-2}\mathcal{S}_{D,1}^{(1)} & -v^{-2}\mathcal{S}_{D,1}^{(1)} \\ v_b^{-2}\mathcal{K}_{D,1}^{(1)} & 0 \end{bmatrix},$$

$$\mathcal{A}_{1,2,0} := \begin{bmatrix} v_b^{-2}(-\ln v_b \mathcal{S}_{D,1}^{(1)} + \mathcal{S}_{D,1}^{(2)}) & -v^{-2}(-\ln v \mathcal{S}_{D,1}^{(1)} + \mathcal{S}_{D,1}^{(2)}) \\ v_b^{-2}(-\ln v_b \mathcal{K}_{D,1}^{(1)} + \mathcal{K}_{D,1}^{(2)}) & 0 \end{bmatrix},$$

and

$$\mathcal{A}_{0,1} := \begin{bmatrix} 0 & 0 \\ 0 & -(\frac{1}{2}I + \mathcal{K}_D^*) \end{bmatrix}.$$

The above operators are defined, for $\phi \in L^2(\partial D)$, as

$$\mathcal{S}_D[\phi](x) := \frac{1}{2\pi} \int_{\partial D} \ln |x - y| \phi(y)\, d\sigma(y),$$

$$\hat{\mathcal{S}}_D^k[\phi](x) := \mathcal{S}_D[\phi](x) + \eta_k \int_{\partial D} \phi\, d\sigma, \quad \eta_k := \frac{1}{2\pi}(\ln k + \gamma - \ln 2) - \frac{i}{4},$$

$$\mathcal{S}_{D,1}^{(1)}[\phi](x) := \int_{\partial D} b_1 |x - y|^2 \phi(y)\, d\sigma(y),$$

$$\mathcal{S}_{D,1}^{(2)}[\phi](x) := \int_{\partial D} b_1 |x - y|^2 \ln |x - y| \phi(y) + c_1 |x - y|^2 \phi(y)\, d\sigma(y),$$

$$\mathcal{K}_{D,1}^{(1)}[\phi](x) := \int_{\partial D} b_1 \frac{\partial |x - y|^2}{\partial \nu(x)} \phi(y)\, d\sigma(y),$$

$$\mathcal{K}_{D,1}^{(2)}[\phi](x) := \int_{\partial D} b_1 \frac{\partial |x - y|^2 \ln |x - y|}{\partial \nu(x)} \phi(y) + c_1 \frac{\partial |x - y|^2}{\partial \nu(x)} \phi(y)\, d\sigma(y),$$

where

$$b_1 := -\frac{1}{8\pi} \quad \textit{and} \quad c_1 := -\frac{1}{8\pi}\left(\gamma - \ln 2 - 1 - \frac{i\pi}{2}\right), \tag{3.18}$$

and $\gamma = 0.5772\ldots$ *is the Euler constant.*

The operator $\mathcal{S}_D$ is the Laplace single layer potential associated with D. In Chapter 4, when we study the analogous three-dimensional problem, this low-frequency approach will be particularly helpful, as the leading-order operator $\mathcal{S}_D$ is invertible. Since we are presently working in two dimensions $\mathcal{S}_D$ is not generally invertible, however the following two results help us understand the extent of its degeneracy.

Lemma 3.2.2. *If for some* $\phi \in L^2(\partial D)$ *with* $\int_{\partial D} \phi = 0$, *it holds that* $\mathcal{S}_D[\phi](x) = 0$ *for all* $x \in \partial D$, *then* $\phi = 0$ *on* ∂D.

Proof. The arguments given in Lemma 2.25 of [17] can be easily generalized to the case where D is the disjoint union of a finite number of bounded Lipschitz domains in $\mathbb{R}^2$. □

Lemma 3.2.3. *Independent of the number* $N \in \mathbb{N}$ *of connected components making up* D, *we have that*

$$\dim \ker \mathcal{S}_D \leq 1.$$

Proof. Let $\psi \in \ker \mathcal{S}_D$. Thanks to Lemma 3.2.2, if $\int_{\partial D} \psi = 0$ then $\psi = 0$. Suppose that $\int_{\partial D} \psi \neq 0$ and take some other $\tilde{\psi} \in \ker \mathcal{S}_D$ with $\int_{\partial D} \tilde{\psi} \neq 0$. If we define the function

$$f = \frac{\psi}{\int_{\partial D} \psi} - \frac{\tilde{\psi}}{\int_{\partial D} \tilde{\psi}},$$

then f satisfies $\mathcal{S}_D[f] = 0$ and $\int_{\partial D} f = 0$ so by Lemma 3.2.2 we have that $f = 0$. Therefore $\psi = (\int_{\partial D} \psi / \int_{\partial D} \tilde{\psi})\tilde{\psi}$. □

There are two cases to consider, in light of Lemma 3.2.3:
- Case I: $\dim \ker \mathcal{S}_D = 1$;
- Case II: $\dim \ker \mathcal{S}_D = 0$.

By the Fredholm Alternative theorem, an equivalent formulation is
- Case I: $\mathcal{S}_D$ is not invertible;
- Case II: $\mathcal{S}_D$ is invertible;

as an operator from $L^2(\partial D)$ to $H^1(\partial D)$. We are able to prove an important property of the operator $\hat{\mathcal{S}}_D^k$ that was defined in Lemma 3.2.1 and is the leading-order approximation to $\mathcal{S}_D^k$.

Lemma 3.2.4. *Given $k \in \mathbb{C} \setminus \{z \in \mathbb{C} : \Re(z) = 0, \Im(z) \geq 0\}$, the operator $\hat{\mathcal{S}}_D^k$ is invertible in the space of bounded linear functions $\mathcal{L}(L^2(\partial D), H^1(\partial D))$.*

Proof. Since $\hat{\mathcal{S}}_D^k$ is Fredholm with index 0, we need only to show that it is injective. To this end, assume that $y \in L^2(\partial D)$ is such that

$$\hat{\mathcal{S}}_D^k[y] = \mathcal{S}_D[y] + \eta_k \int_{\partial D} y = 0. \tag{3.19}$$

Case I: Let ψ_0 be the unique element of $\ker \mathcal{S}_D$ with $\int_{\partial D} \psi_0 = 1$ (which exists as a result of Lemma 3.2.2). We then find that $\mathcal{S}_D[y] \perp \psi_0$ in $L^2(\partial D)$ and hence (3.19) becomes

$$\eta_k \left(\int_{\partial D} y \right) \left(\int_{\partial D} \psi_0 \right) = 0.$$

Thus $\int_{\partial D} y = 0$. It follows from (3.19) that $\mathcal{S}_D[y] = 0$ and further by Lemma 3.2.2 we have that $y = 0$.

Case II: Define $\psi_0 = \mathcal{S}_D^{-1}[\chi_{\partial D}]$, then (3.19) gives us that

$$\mathcal{S}_D[y] = -\eta_k \int_{\partial D} y,$$

is constant so, since $\mathcal{S}_D$ is injective, we find that $y = c\psi_0$ for some c. Substituting back into (3.19) gives

$$c\left(1 + \eta_k \int_{\partial D} \psi_0 \right) = 0.$$

Everything within the brackets is real with the one exception of η_k (which has non-zero imaginary part, thanks to the choice of k) so we must have that $c = 0$. □

We are now ready to compute the resonant frequencies and associated eigenmodes for our system of subwavelength resonators. Manipulating the first entry of (3.13), gives us that

$$\hat{\mathcal{S}}_D^{k_b}[\phi] - \hat{\mathcal{S}}_D^k[\psi] = \hat{\mathcal{S}}_D^k[\phi - \psi] + \frac{1}{2\pi} \ln \frac{v}{v_b} \int_{\partial D} \phi,$$

hence it holds that

$$\psi = \phi + \frac{1}{2\pi} \ln \frac{v}{v_b} \left(\int_{\partial D} \phi \right) (\hat{\mathcal{S}}_D^k)^{-1}[\chi_{\partial D}] + O(\omega^2). \tag{3.20}$$

Here, $\chi_{\partial D}$ is used to denote the characteristic function of ∂D.

To deal with the second component of (3.13) we first prove some technical lemmas.

Lemma 3.2.5. *For any $\phi \in L^2(\partial D)$ and $j = 1, \dots, N$, we have that*
(i) $\int_{\partial D_j} (\frac{1}{2}I - \mathcal{K}_D^*)[\phi] = 0$,
(ii) $\int_{\partial D_j} (\frac{1}{2}I + \mathcal{K}_D^*)[\phi] = \int_{\partial D_j} \phi$.

Proof. (i) Since $\mathcal{S}_D[\phi]$ is harmonic in D,

$$\int_{\partial D_j} \left(\frac{1}{2}I - \mathcal{K}_D^*\right)[\phi] = -\int_{\partial D_j} \frac{\partial}{\partial \nu} \mathcal{S}_D[\phi]|_- = -\int_{D_j} \Delta \mathcal{S}_D[\phi] = 0.$$

Then (ii) is immediate. □

Lemma 3.2.6. *For any $\phi \in L^2(\partial D)$ and $j = 1, \dots, N$, we have that*
(i) $\int_{\partial D_j} \mathcal{K}_{D,1}^{(1)}[\phi] = 4b_1 |D_j| \int_{\partial D} \phi$,
(ii) $\int_{\partial D_j} \mathcal{K}_{D,1}^{(2)}[\phi] = -\int_{D_j} \mathcal{S}_D[\phi] + (4b_1 + 4c_1)|D_j| \int_{\partial D} \phi$,

where $|D_i|$ is the area of D_i and b_1 and c_1 are the constants defined in (3.18).

Proof. (i) follows from the divergence theorem:

$$\begin{aligned}\int_{\partial D_j} \mathcal{K}_{D,1}^{(1)}[\phi](x)\, d\sigma(x) &= b_1 \int_{D_j} \int_{\partial D} \Delta_x |x-y|^2 \phi(y)\, d\sigma(y)\, dx \\ &= 4b_1 |D_j| \int_{\partial D} \phi(y)\, d\sigma(y).\end{aligned}$$

Similarly for (ii), we can show that

$$\begin{aligned}\int_{\partial D_j} K_{D,1}^{(2)}[\phi](x)\, d\sigma(x) &= \int_{D_j} \int_{\partial D} \Delta_x [|x-y|^2 (b_1 \ln|x-y| + c_1)] \phi(y)\, d\sigma(y)\, dx \\ &= -\int_{D_j} \mathcal{S}_D[\phi](x)\, dx + (4b_1 + 4c_1)|D_j| \int_{\partial D} \phi(y)\, d\sigma(y),\end{aligned}$$

making use of the fact that $b_1 = -1/8\pi$. □

Turning now to the second component of (3.13), we see that

$$\begin{aligned}&\left(-\frac{1}{2}I + \mathcal{K}_D^* + v_b^{-2} \mathcal{K}_{D,1}^{(1)} \omega^2 \ln \omega + v_b^{-2}(-\ln v_b \mathcal{K}_{D,1}^{(1)} + \mathcal{S}_{D,1}^{(2)})\omega^2\right)[\phi] \\ &\quad - \delta\left(\frac{1}{2}I + \mathcal{K}_D^*\right)[\psi] = O(\delta \omega^2 \ln \omega) + O(\omega^4 \ln \omega).\end{aligned}$$

We substitute expression (3.20) for ψ to see that ϕ satisfies the equation

$$\left(-\frac{1}{2}I+\mathcal{K}_D^*\right)[\phi]+(v_b^{-2}\mathcal{K}_{D,1}^{(1)}\omega^2\ln\omega+v_b^{-2}(-\ln v_b\mathcal{K}_{D,1}^{(1)}+\mathcal{K}_{D,1}^{(2)})\omega^2)[\phi]$$
$$-\delta\left(\frac{1}{2}I+\mathcal{K}_D^*\right)[\phi]-\frac{1}{2\pi}\delta\ln\frac{v}{v_b}\left(\int_{\partial D}\phi\right)\left(\frac{1}{2}I+\mathcal{K}_D^*\right)[(\hat{\mathcal{S}}_D^k)^{-1}[\chi_{\partial D}]]$$
$$=O(\delta\omega^2\ln\omega)+O(\omega^4\ln\omega). \tag{3.21}$$

At leading order (3.21) is just $(-\frac{1}{2}I+\mathcal{K}_D^*)[\phi]=0$. This kernel can be characterized by the following two lemmas.

Lemma 3.2.7. *If $\phi\in L^2(\partial D)$ is such that $\phi\in\ker(-\frac{1}{2}I+\mathcal{K}_D^*)$, then there exist constants b_j such that $\mathcal{S}_D[\phi]=\sum_{j=1}^N b_j\chi_{\partial D_j}$.*

Proof. Let $u:=\mathcal{S}_D[\phi]$. Then $\Delta u=0$ in D and $\frac{\partial u}{\partial\nu}|_-=(-\frac{1}{2}I+\mathcal{K}_D^*)[\phi]=0$ on ∂D so u satisfies a homogeneous interior Neumann problem on each of the N connected components $D_1,\dots,D_N$ of D. It is known that such problems are uniquely solvable up to the addition of a constant. □

Lemma 3.2.8. *Let $k_0\in\mathbb{C}\setminus\{0\}$. The set of vectors $\{\psi_1,\dots,\psi_N\}$ defined as*

$$\psi_i:=(\hat{\mathcal{S}}_D^{k_0})^{-1}[\chi_{\partial D_i}], \tag{3.22}$$

forms a basis for the space $\ker(-\frac{1}{2}I+\mathcal{K}_D^*)$.

Proof. The linear independence of $\{\psi_1,\dots,\psi_N\}$ follows from the linearity and injectivity of $\hat{\mathcal{S}}_D^{k_0}$, plus the independence of $\{\chi_{\partial D_1},\dots,\chi_{\partial D_N}\}$.

For $\phi\in L^2(\partial D)$, the difference between $\hat{\mathcal{S}}_D^{k_0}[\phi](x)$ and $\mathcal{S}_D[\phi](x)$ is a constant (in x) so they will have the same derivatives. In particular, they are both harmonic and satisfy the same jump conditions across ∂D. Therefore, using arguments as in Lemma 3.2.7, we see that if $\phi\in\ker(-\frac{1}{2}I+\mathcal{K}_D^*)$ then $\hat{\mathcal{S}}_D^{k_0}[\phi]\in\operatorname{span}\{\chi_{\partial D_1},\dots,\chi_{\partial D_N}\}$. Thus $\phi\in\operatorname{span}\{\psi_1,\dots,\psi_N\}$. □

From Lemma 3.2.8, we know that $\ker(-\frac{1}{2}I+\mathcal{K}_D^*)$ has dimension equal to the number of connected components of D (a wider discussion can be found in, e. g., [3]). Thus we can take a basis

$$\{\phi_1,\dots,\phi_N\},$$

of the null space $\ker(-\frac{1}{2}I+\mathcal{K}_D^*)$. Then, in light of the fact that at leading order (3.21) is just $(-\frac{1}{2}I+\mathcal{K}_D^*)[\phi]=0$, it is natural to seek a solution of the form

$$\phi=\sum_{j=1}^N a_j\phi_j+O(\omega^2\ln\omega+\delta), \tag{3.23}$$

for some nontrivial constants a_j. The solutions (ϕ,ψ) to (3.13) are determined only up to multiplication by a constant (and hence so are $a_1,\dots,a_N$). We fix the scaling to be such that the eigenmodes are normalized in the $L^2(D)$-norm.

We now integrate (3.21) over each $\partial D_i, i = 1, \ldots, N$ and use the results of Lemmas 3.2.5 and 3.2.6 to find that, up to an error of $O(\delta\omega^2 \ln \omega) + O(\omega^4 \ln \omega)$,

$$B_\delta^{(i)}(\omega)[\phi] := \Big(\int_{\partial D} \phi\Big)\bigg(\omega^2 \ln \omega + \bigg(\Big(1 + \frac{c_1}{b_1} - \ln v_b\Big) - \frac{\mathcal{S}_D[\phi]|_{\partial D_i}}{4b_1(\int_{\partial D} \phi)}\bigg)\omega^2\bigg)$$
$$- \frac{v_b^2}{4b_1|D_i|}\bigg[\int_{\partial D_i} \phi + \frac{\ln(v/v_b)}{2\pi}\Big(\int_{\partial D} \phi\Big) \int_{\partial D_i} (\hat{\mathcal{S}}_D^k)^{-1}[\chi_{\partial D}]\bigg]\delta = 0.$$

When we substitute the expression (3.23) for ϕ we reach the system of equations, up to an error of order $O(\delta\omega^2 \ln \omega) + O(\omega^4 \ln \omega)$,

$$\begin{pmatrix} B_\delta^{(1)}(\omega)[\phi_1] & B_\delta^{(1)}(\omega)[\phi_2] & \cdots & B_\delta^{(1)}(\omega)[\phi_N] \\ \vdots & \vdots & \ddots & \vdots \\ B_\delta^{(N)}(\omega)[\phi_1] & B_\delta^{(N)}(\omega)[\phi_2] & \cdots & B_\delta^{(N)}(\omega)[\phi_N] \end{pmatrix} \begin{pmatrix} a_1 \\ \vdots \\ a_N \end{pmatrix} = 0. \tag{3.24}$$

Thus, our asymptotic analysis has arrived at a method for finding the subwavelength resonant frequencies $\omega = \omega(\delta)$, which are the values at which the matrix in (3.24) is singular. Since the operators $B_\delta^{(i)}$ are linear, the solutions $\omega = \omega(\delta)$ to (3.24) are independent of the choice of basis $\{\phi_1, \ldots, \phi_N\}$. While it is not immediately obvious that this is any more efficient than the numerical scheme proposed below, it has the advantage that the quantities in the definition of $B_\delta^{(i)}(\omega)$ are all relatively straightforward to compute. Conversely, when we come to study three-dimensional systems in Chapter 4, the equivalent version of (3.24) will be much simpler, to the extent that the resonances can be solved simply by finding the eigenvalues of a square matrix.

Another valuable consequence of the analysis presented in this section is that it allows us to prove that a system of N coupled subwavelength resonators has N subwavelength resonant frequencies. This follows from the fact that $\mathcal{A}(0, 0)$ has an N-dimensional kernel, a consequence of Lemma 3.2.8. Subsequently, if δ is positive but small, then the asymptotic perturbation theory of Gohberg and Sigal [83, 14] shows that $\mathcal{A}(\omega(\delta), \delta)$ has N characteristic values (with positive real parts), which are such that $\omega \to 0$ as $\delta \to 0$.

Theorem 3.2.1. *A system of N coupled subwavelength resonators exhibits N subwavelength resonant frequencies with positive real parts, counted up to multiplicity.*

3.3 Numerical methods

The integral formulation (3.11) is useful not only for asymptotic analysis, but also for numerical computation. Indeed, boundary integral formulations such as this are the foundation of the boundary element method (BEM) [129]. These methods are based on partitioning the boundaries of the resonators ∂D into finite-sized elements, in order to obtain a discrete version of the problem, that can be handled numerically. These methods have been developed extensively and can handle many different shapes of domains, with varying degrees of computational expense.

Here, we will summarize one very simple method for obtaining the eigenvalues numerically, which will be sufficient for the examples that we will consider in this book. This is an example of a *multipole expansion method*, which we will present specifically for the case of circular resonators. In the case that each D_n is a circle, the boundary can be described using polar coordinates, in terms of an angle θ. Crucially, the density functions ϕ and ψ, which belong to $L^2(\partial D)$, are both 2π-periodic functions of θ, meaning they can be represented efficiently using Fourier series. This is the main idea behind the approach.

The integral formulation (3.11) can be helpfully restated as

$$\begin{pmatrix} \mathcal{S}_D^{k_b} & -\mathcal{S}_D^{k} \\ \frac{\partial}{\partial \nu}\mathcal{S}_D^{k_b}\big|_- & -\delta \frac{\partial}{\partial \nu}\mathcal{S}_D^{k}\big|_+ \end{pmatrix} \begin{pmatrix} \phi \\ \psi \end{pmatrix} = \begin{pmatrix} 0 \\ 0 \end{pmatrix}, \tag{3.25}$$

where equality holds as elements of $L^2(\partial D)$ and the densities $(\phi, \psi) \in L^2(\partial D) \times L^2(\partial D)$ are the unknowns to be found.

Since we are interested in the case of circular resonators, ϕ and ψ are, on each ∂D_n, $n = 1, \dots, N$, 2π-periodic functions of θ_n where (r_n, θ_n) denotes a polar coordinate system about on the center of D_n. Such functions admit Fourier expansions of the form

$$\phi|_{\partial D_n} = \sum_{m \in \mathbb{Z}} a_m^n e^{im\theta_n}, \tag{3.26}$$

for coefficients a_m^n, and similarly for ψ. The reason such an expansion is useful is that $\mathcal{S}_{D_n}^k[e^{im\theta_n}]$ has an explicit representation, shown in [15] to be given by

$$\mathcal{S}_{D_n}^k[e^{im\theta_n}] = \begin{cases} c_n J_m(kR_n) H_m^{(1)}(kr_n) e^{im\theta_n}, & r_n > R_n, \\ c_n H_m^{(1)}(kR_n) J_m(kr_n) e^{im\theta_n}, & r_n \le R_n, \end{cases} \tag{3.27}$$

where J_m and $H_m^{(1)}$ are the Bessel and Hankel functions of the first kind, respectively, $c_n = -\frac{i\pi R_n}{2}$ and R_n is the radius of D_n.

In order to apply this method to the case of $N \in \mathbb{N}$ resonators we also require an expression for $\mathcal{S}_{D_{n'}}^k[e^{im\theta_n}]$, where $n \neq n'$. This is achieved through the use of Graf's addition formula [112], which says that for any $x, y \in \mathbb{R}^2$ such that $|x| > |y|$ the Helmholtz fundamental solution G^k can be expanded as

$$G^k(x - y) = -\frac{i}{4} \sum_{l \in \mathbb{Z}} H_l^{(1)}(k|x|) J_l(k|y|) e^{il(\vartheta_x - \vartheta_y)}, \tag{3.28}$$

where $x = (|x|, \vartheta_x)$ and $y = (|y|, \vartheta_y)$ are polar representations around a common origin.

Finally, we make the identification $L^2(\partial D) \cong L^2(\partial D_1) \times \cdots \times L^2(\partial D_N)$ and decompose the single layer potential as

$$\mathcal{S}_D^k = \begin{pmatrix} \mathcal{S}_{D_1}^k & \mathcal{S}_{D_2}^k|_{\partial D_1} & \cdots & \mathcal{S}_{D_N}^k|_{\partial D_1} \\ \mathcal{S}_{D_1}^k|_{\partial D_2} & \mathcal{S}_{D_2}^k & \cdots & \mathcal{S}_{D_N}^k|_{\partial D_2} \\ \vdots & \vdots & \ddots & \vdots \\ \mathcal{S}_{D_1}^k|_{\partial D_N} & \mathcal{S}_{D_2}^k|_{\partial D_N} & \cdots & \mathcal{S}_{D_N}^k \end{pmatrix}, \tag{3.29}$$

where $\mathcal{S}_{D_{n'}}^k|_{\partial D_n} : L^2(\partial D_{n'}) \to L^2(\partial D_n)$ is the evaluation of $\mathcal{S}_{D_{n'}}^k$ on ∂D_n. Let $z_n^{n'}$ be the vector from the center of D_n to that of $D_{n'}$, then the off-diagonal terms in (3.29) are of the form $\mathcal{S}_{D_{n'}}^k[e^{im\theta_n}] = \mathcal{S}_{z_n^{n'}+D_n}^k[e^{im\theta_n}]$. The addition of $z_n^{n'}$ within the integrand can then be decomposed using (3.28).

The derivatives appearing in (3.25) can be handled similarly, based on the expressions

$$\partial_\nu^-[\mathcal{S}_{D_n}^k[e^{im\theta_n}]] = c_n k J_m'(kR_n) H_m^{(1)}(kR_n) e^{im\theta_n}, \tag{3.30}$$

$$\partial_\nu^+[\mathcal{S}_{D_n}^k[e^{im\theta_n}]] = c_n k J_m(kR_n) H_m^{(1)\prime}(kR_n) e^{im\theta_n}, \tag{3.31}$$

which can be derived by differentiating (3.27).

Finally, we truncate the Fourier basis on each ∂D_n, using only $\{e^{im\theta_n} : n = -M, \ldots, M\}$ for some $M \geq 0$, to reach an approximate matrix representation for (3.25). Given the matrix representation, numerical root finding can be used to find the resonant frequencies. That is, given the matrix $A(\omega)$, which is a discrete approximation of the operator $\mathcal{A}(\omega, \delta)$ for fixed δ, we wish to find the roots of the function

$$f(\omega) = \min\left|\sigma(A(\omega))\right|. \tag{3.32}$$

That is, f is the (absolute value of) the minimum eigenvalue of the matrix $A(\omega)$.

Our preferred numerical root finding approach, which is sufficient for this example, is Muller's method. This is an extension of the secant method and is an efficient and reliable interpolation method for finding a zero of a function defined on the complex plane. It has the advantage over Newton's method that the derivatives of the function need not be computed, which is helpful in our case.

The secant method performs root finding by taking two points on the graph of a function f and finding an approximate root by approximating the function as being linear between these two points. Muller's method builds on this by approximating the function with a quadratic. For this to be achieved, we need to fit the quadratic to *three* points on the graph of f.

We need to find the quadratic $Q_f(\omega)$ that passes through the points $(\omega_0, f(\omega_0))$, $(\omega_1, f(\omega_1))$, and $(\omega_2, f(\omega_2))$. Suppose that $Q_f(\omega)$ is given by

$$Q_f(\omega) = a(\omega - \omega_2)^2 + b(\omega - \omega_2) + c,$$

for some constants a, b, and c, which we need to find. We have that

$$\begin{aligned} f(\omega_0) &= a(\omega_0 - \omega_2)^2 + b(\omega_0 - \omega_2) + c, \\ f(\omega_1) &= a(\omega_1 - \omega_2)^2 + b(\omega_1 - \omega_2) + c, \\ f(\omega_2) &= a(\omega_2 - \omega_2)^2 + b(\omega_2 - \omega_2) + c. \end{aligned}$$

Solving for a, b, and c we obtain

$$\begin{aligned} a &= \frac{(\omega_1 - \omega_2)(f(\omega_0) - f(\omega_2)) - (\omega_0 - \omega_2)(f(\omega_1) - f(\omega_2))}{(\omega_0 - \omega_1)(\omega_0 - \omega_2)(\omega_1 - \omega_2)}, \\ b &= \frac{(\omega_0 - \omega_2)^2(f(\omega_1) - f(\omega_2)) - (\omega_1 - \omega_2)^2(f(\omega_0) - f(\omega_2))}{(\omega_0 - \omega_1)(\omega_0 - \omega_2)(\omega_1 - \omega_2)}, \\ c &= f(\omega_2). \end{aligned}$$

Once we have the coefficients a, b, and c, we can find the root. Using the quadratic formula, we find that the root ω_3 is given by

$$\omega_3 = \omega_2 - \frac{2c}{b \pm \sqrt{b^2 - 4ac}},$$

where the sign of the square root should be chosen so as to maximize the (absolute value of) the denominator. This version of the quadratic formula has been chosen to improve numerical stability.

This procedure can be repeated to find ω_4, ω_5, and so on. It can be terminated when $|f(\omega_n)|\tau_f$ and $|\omega_n - \omega_{n-1}| < \tau_\omega$, where τ_f and τ_ω are some given tolerances. It can be shown that Muller's method converges to a single root at an order of approximately 1.84 (given exactly as the largest root of the equation $\zeta^3 - \zeta^2 - \zeta - 1 = 0$). This demonstrates that there is only a slight loss of convergence speed compared to the quadratic convergence of Newton's method, for example, with the significant benefit of not needing to compute any derivatives of f.

We can use this approach to find, for each fixed $\delta > 0$, the N values of $\omega \in \mathbb{C}$ such that there exists a nontrivial solution to (3.24). As an example, for the case where $N = 50$ the results are shown in Figure 3.2. We show the resonant frequencies with positive real parts, but have excluded the first resonance $\omega_1 = 0.0002284 - 5.26 \times 10^{-5}\mathrm{i}$ for the sake of convenience (its imaginary part is much larger than the others).

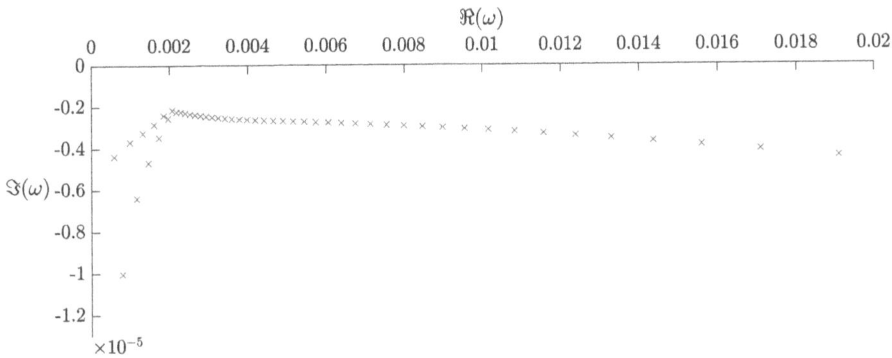

Figure 3.2: The subwavelength resonant frequencies, plotted in the complex plane, of a system of 50 identical resonators arranged linearly with each being 1.05 times the size of the previous. The first resonance $\omega_1 = 0.0002284 - 5.26 \times 10^{-5}i$ is omitted for convenience, since its imaginary part is much larger than the others.

3.4 Cochlea-inspired graded metasurface

Now that we have developed suitable asymptotic and numerical methods, we will consider the design of a graded system of subwavelength resonators. We will design this so that it replicates the frequency separation that takes place in the cochlea. In the middle of the twentieth century, Georg von Békésy performed a series of famous experiments that revealed new insight into how the cochlea functions [161]. One of his main discoveries was the existence of a relationship between incoming frequency and the position in the cochlea where the sound is most strongly detected. This *tonotopic map* is perhaps the most fundamental property of cochlear function. His results showed that the frequency $f(x)$ giving rise to maximum excitation at a distance x from the base of the cochlea satisfies a tonotopic map of the form

$$f(x) = ae^{-x/d} + c, \tag{3.33}$$

for some $a, d, c \in \mathbb{R}$ [161].

The exponential form of the desired tonotopic map (3.33) immediately suggests that we should grade the resonator's parameters accordingly. As a demonstrative example, we elect to do so by keeping the spacing of the resonators fixed and increasing their radius as an exponential function of their position in the array. For an array of 50 such resonators, the (absolute value of the) eigenmodes are shown in Figures 3.3b–d. Each mode features oscillations until a clear peak that is followed by a rapid decrease in amplitude, as is typical of rainbow trapping phenomena in graded metamaterials.

In Figure 3.3e, we show the relationship between the position of maximum amplitude of each eigenmode and the associated resonant frequency. We see that, if some of the lowest frequency modes are ignored, the pattern follows a relationship that is approximately of the form (3.33). The dashed line shows a relationship of the form (3.33).

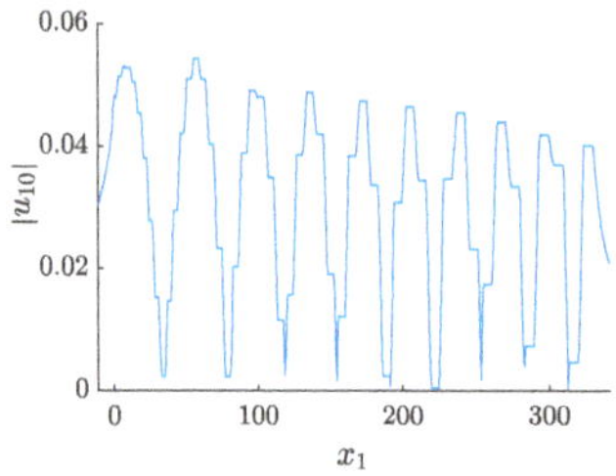

(a) $\omega_{10} = 0.003803 - 0.000003i$

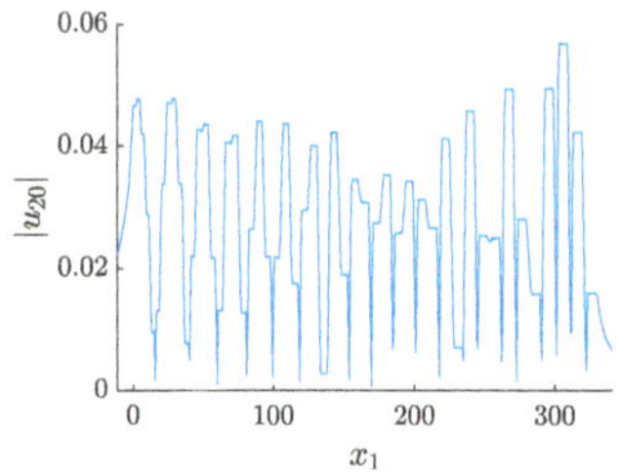

(b) $\omega_{20} = 0.006122 - 0.000001i$

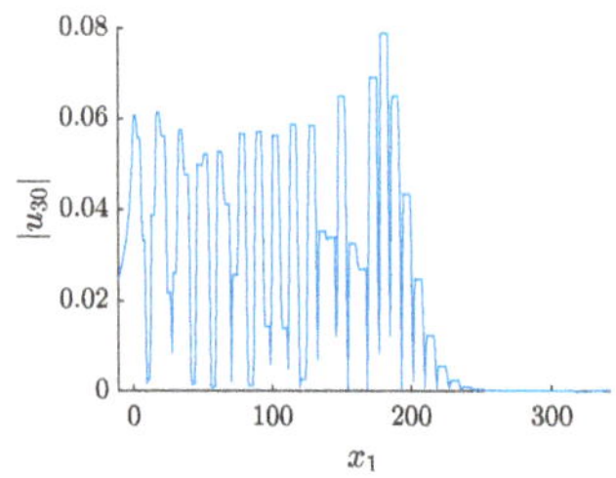

(c) $\omega_{30} = 0.007612 - 0.000001i$

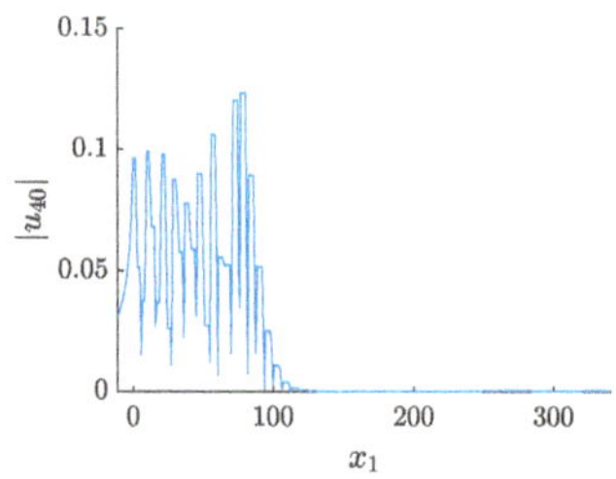

(d) $\omega_{40} = 0.01049 - 0.000002i$

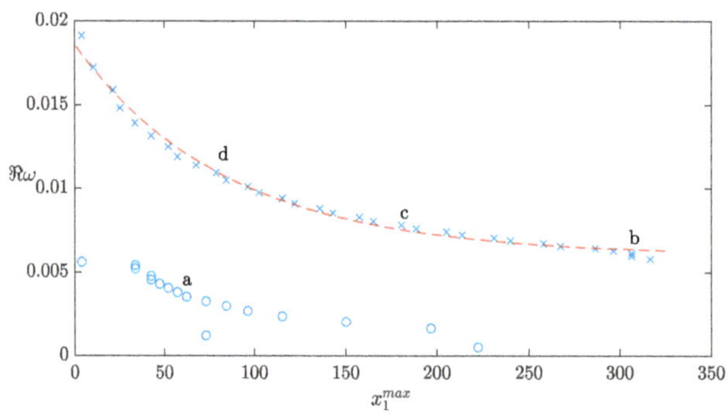

(e) Tonotopic map: the (real part of the) resonant frequency against the position of maximum amplitude x_1^{max}.

Figure 3.3: The existence of a tonotopic map for a passive system of graded subwavelength resonators. (e) shows, for each eigenmode, the relationship between the real part of the associated resonant frequency $\Re\omega$ and the location (x_1-coordinate) of the maximum amplitude. We study the case of 50 resonators. A (least squares) approximation to the relationship exhibited by the crosses is shown, this has equation $0.0126e^{-0.0117x} + 0.0060$. The 17 circles are excluded from this calculation. Subplots (a)–(d) show the eigenmodes corresponding to the points marked on the top plot. We depict the amplitude of each eigenmode $|u_n| = |u_n(x_1, 0)|$ along the line $x_2 = 0$ (through the centers of the resonators).

This was obtained by fitting a curve of this form in a least-squares sense and has parameter values $a = 0.0126$, $d = -0.0117$ and $c = 0.0060$ (in this dimensionless case). Figure 3.3a shows an example of a mode that is never cut off since its eigenfrequency is below the bottom of the lowest position of the effective band gap. These modes (indicated by circular markers in Figure 3.3e) were excluded from the least-squares curve fitting. In order to also reflect these frequencies, we would need to increase the slope of the gradient so that the largest resonators are even larger. However, in doing so we would start to bring the resonators into a close-to-touching regime, which is known to lead to blow up of the gradient of the field in the small gap, thereby posing significant challenges for our numerical methods [10].

From Figures 3.3a–d, it is notable that the solution is approximately constant on each resonator. This is because the solution, taking the form (3.9), is given by $\hat{\mathcal{S}}_D^{k_b}[\phi]$ at leading order, which by Lemma 3.2.7 is constant for $\phi \in \ker(-\frac{1}{2}I + \mathcal{K}_D^*)$. The physical interpretation of this is that the resonators act as monopolar scatterers.

The system sketched in Figures 3.2 and 3.3 has arbitrary dimensions, meaning that the frequency values are nonphysical. However, the main advantage of using the high-contrast subwavelength resonators that we have chosen here is that they are able to perform the desired spatial frequency separation at the same scale as the cochlea. Taking the material parameters of air and water in the resonators and background medium, for the sake of having a physically meaningful example, we can develop a design that has similar dimensions to the cochlea. In Figure 3.4 we show the 22 subwavelength resonant modes and associated resonant frequencies for a system of 22 cylindrical resonators that are graded in size. The array is setup so that the first resonator has radius $R_1 = 0.1\,\mathrm{mm}$ and each successive resonator is 1.05 times larger than the previous. This structure has a total length of 35 mm, similar to the cochlea, and its resonant frequencies have real parts which fall within the range of audible frequencies (often quoted as 20 Hz – 20 kHz).

Of course, the structure will also have higher order resonant modes at frequencies which correspond to wavelengths similar to the size of the resonators, or bigger. These modes will not be described by our asymptotic methods. However, given the physical dimensions and wavelengths of the problem we are interested in, we can focus our attention on the subwavelength modes as these will dominate the behavior of the system.

3.5 Cochlear membrane modes

As a brief aside, it is valuable to pause and compare the properties of the eigenmodes in Figure 3.4 with those of the cochlear membrane. The N subwavelength eigenmodes take the form of increasingly oscillating coupled patterns. The profiles share several similarities with the response of the cochlear partition, some examples of which are given in Figure 3.5. These were computed using a simple ordinary differential equation (ODE) model proposed by Duke and Jülicher [71]. In this model, wave propagation problem in the frequency domain is modeled by the equation

$$0 = \frac{2\rho\omega^2}{l}\tilde{p} + K(\omega)\frac{\partial^2 \tilde{p}}{\partial x^2}, \quad x \in [0, L],\ \omega \in \mathbb{R}, \tag{3.34}$$

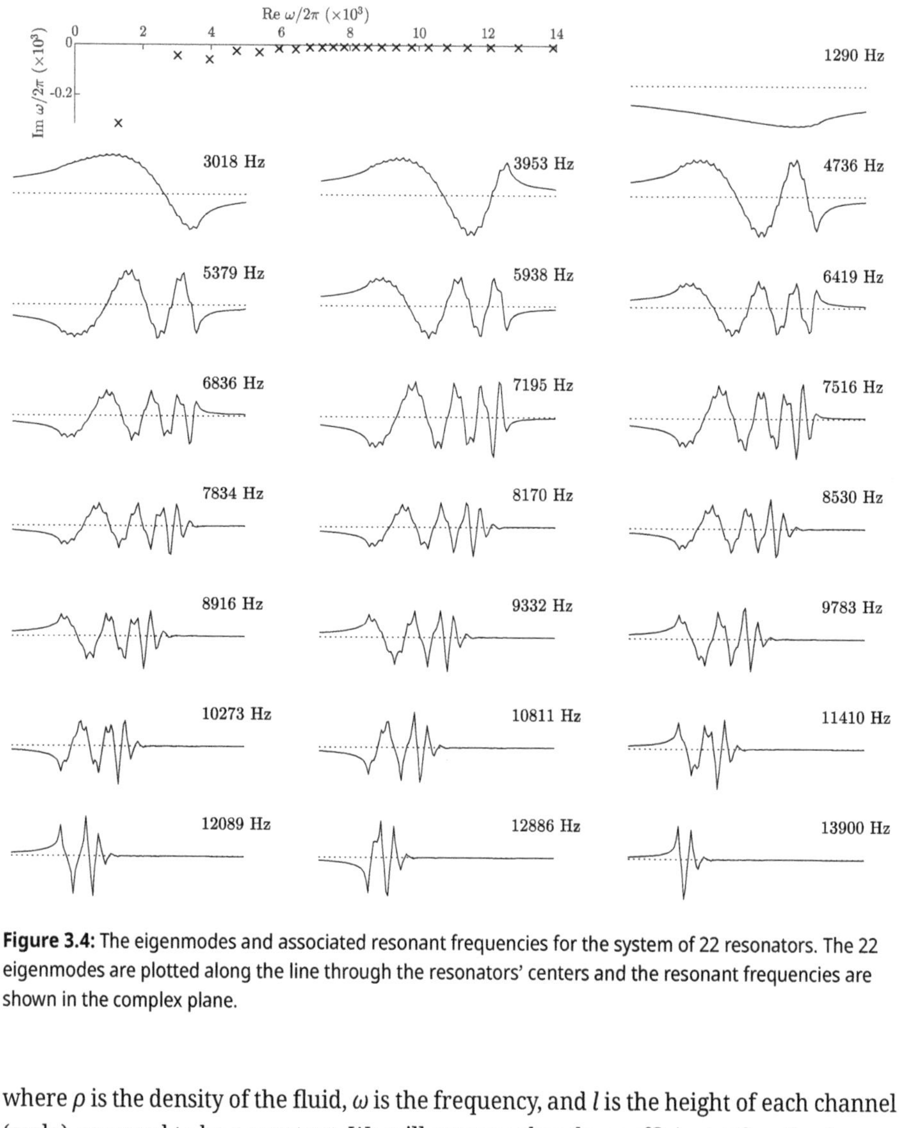

Figure 3.4: The eigenmodes and associated resonant frequencies for the system of 22 resonators. The 22 eigenmodes are plotted along the line through the resonators' centers and the resonant frequencies are shown in the complex plane.

where ρ is the density of the fluid, ω is the frequency, and l is the height of each channel (scala), assumed to be a constant. We will suppose that the coefficient K has the form

$$K(x,\omega) = \alpha\big(k(x) - \omega\big), \tag{3.35}$$

where the function $\alpha k(x)$ is the passive stiffness of the membrane at $x \in [0, L]$ and $\alpha \in \mathbb{R}$ is a constant.

We solve (3.34) by integrating from $x = L$ towards $x = 0$. We enforce the boundary conditions $\tilde{p}(L) = 0$ (to represent the pressure equality at the helicotrema, where the two channels meet) and the normalized condition $\partial_x \tilde{p}(L) = 1$. We use $\rho = 10^3\,\mathrm{kg\,m^{-3}}$ as above and adopt the values $L = 35\,\mathrm{mm}$, $l = 1\,\mathrm{mm}$, and $\alpha = 10^4\,\mathrm{Pa\,m^{-1}\,s^{-1}}$ from [71].

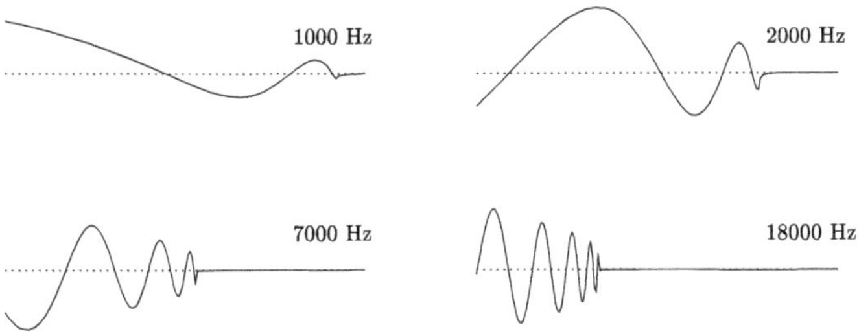

Figure 3.5: The eigenmodes of the passive cochlear partition, for a range of audible frequencies, calculated using the approach of Duke and Jülicher [71].

Further, since the frequency–position (tonotopic) map in the cochlea is well known [161], we can infer that k should have the form

$$k(x) = k_0 e^{-x/d}, \tag{3.36}$$

where $k_0 = 10^5\,\mathrm{s}^{-1}$ and $d = 7\,\mathrm{mm}$ [71]. With this stiffness map, the profiles shown in Figure 3.5 are generated.

The key similarities to notice between the membrane modes in Figure 3.5 and the metamaterial resonant modes in Figure 3.4 are that each mode has a position of maximal amplitude (sometimes known as the *resonant place* or the *cut off*) that depends on its resonant frequency [161, 4]. In the region of the resonant place, the wavelength decreases as the amplitude peaks, before the amplitude decays quickly. This was similarly observed by [146] in simulations using point scatterers. The use of resonators with nonzero radii, and the fact that each mode is approximately constant on each resonator, means the profiles look less smooth here.

3.6 Filtering waves

In order to understand how our cochlea-inspired metamaterial acts as a signal processing device, it is useful to consider how the resonator array decomposes an incoming acoustic signal. In particular, in light of the high-contrast system's strongly resonant response, we will be interested in the system's ability to decompose a signal over the subwavelength resonant modes. This idea will be the main focus of Chapter 5, however, it is helpful to briefly consider some of its implications for the metasurface designed here.

Although we will generally study problems in the frequency domain, corresponding to the propagation of time-harmonic waves, it is important to consider how general signals will be scattered by D. Given an incident acoustic pressure wave $p^{\mathrm{in}}(x,t)$, the scattered field $p(x,t)$ will satisfy the scalar wave equation in the interior and exterior of D:

$$\begin{cases} (\nabla \cdot \frac{1}{\rho} \nabla - \frac{1}{\kappa} \frac{\partial^2}{\partial t^2}) p = 0 & \text{for } (x, t) \in \mathbb{R}^2 \setminus \overline{D} \times \mathbb{R}, \\ (\nabla \cdot \frac{1}{\rho_b} \nabla - \frac{1}{\kappa_b} \frac{\partial^2}{\partial t^2}) p = 0 & \text{for } (x, t) \in D \times \mathbb{R}, \end{cases} \tag{3.37}$$

along with appropriate continuity conditions on ∂D and initial conditions.

We wish to briefly offer an explanation of how, given an incident wave $p^{\text{in}}(x, t)$, our system of coupled resonators is able to classify (and hence identify) the sound.

Given an incoming wave, the system of resonators D is able to decompose the signal over its resonant modes. In particular, the subwavelength (low-frequency) part of the signal is decomposed over the subwavelength resonant modes. It is clear that the N subwavelength eigenmodes are linearly independent, giving an N-dimensional space onto which the system projects an incoming signal.

We can approximate the solution by making a time-harmonic decomposition over the eigenmodes in the frequency domain, which we are able to truncate since we are only interested in the response to incident signals corresponding to subwavelength frequencies. The fact that, for $n = 1, \ldots, N$, the Fourier transform of $e^{-\mathrm{i}\omega_n t}$ for $t > 0$ is given by $\mathrm{i}/(\omega - \omega_n)$ motivates the ansatz

$$u(x, \omega) \simeq \sum_{n=1}^{N} \frac{\alpha_n(\omega)\mathrm{i}}{\omega - \omega_n} u_n(x), \tag{3.38}$$

where $\alpha_1, \ldots, \alpha_N$ are complex-valued functions. We will prove a decomposition of this form, from first principles, for the case of three-dimensional resonators in Chapter 5.

The cochlea is able to sense incoming sounds thanks to the receptor cells that it has distributed along its length. Hence, an interesting question is whether knowing the value of the solution on each resonator (which is analogous to the information that the cochlea is able to capture) means that one can recover the functions $\alpha_1, \ldots, \alpha_N$ in (3.38). In practice, this can be done relatively straightforwardly thanks to the fact that the eigenmodes $u_1, \ldots, u_N$ are *nearly* orthogonal in $L^2(D)$. For example, the normalised eigenmodes shown in Figure 3.3 satisfy $(u_n, u_m)_{L^2(D)} = O(10^{-3})$ for $n \neq m$. Hence, we have the following lemma, which will allow us to perform the desired calculations (see also [4] for a proof using the Gram–Schmidt orthonormalization procedure).

Lemma 3.6.1. *Let $\{\omega_1, \ldots, \omega_N\}$ be the subwavelength resonant frequencies of the system $D = D_1 \cup \cdots \cup D_N$ and denote by $u_1, \ldots, u_N$ the corresponding eigenmodes. Then the matrix $\gamma \in \mathbb{C}^{N \times N}$ defined by $\gamma_{ij} = (u_i, u_j)_{L^2(D)}$ is invertible.*

Thanks to this analysis, to find the weight functions $\alpha_1, \ldots, \alpha_N$ in (3.38) we can take the $L^2(D)$-product with $u_n(x)$ for $n = 1, \ldots, N$ and then invert γ. This gives that

$$\begin{pmatrix} \frac{\alpha_1(\omega)\mathrm{i}}{\omega - \omega_1} \\ \vdots \\ \frac{\alpha_N(\omega)\mathrm{i}}{\omega - \omega_N} \end{pmatrix} = \overline{\gamma}^{-1} \begin{pmatrix} (u(\cdot, \omega), u_1)_{L^2(D)} \\ \vdots \\ (u(\cdot, \omega), u_N)_{L^2(D)} \end{pmatrix}. \tag{3.39}$$

Thanks to its representation (3.9) in terms of single layer potentials, $u(\cdot,\omega)$ is a meromorphic function of $\omega \in \mathbb{C}$. Thus, from (3.39) we can see that $\alpha_1,\dots,\alpha_N$ are analytic, and hence we can recover a similar decomposition for $p(x,t)$ using the Laplace inversion theorem (somewhat formally) to see that

$$\begin{aligned} p(x,t) &\simeq \frac{1}{2\pi}\sum_{n=1}^{N} u_n(x)\int_{-\infty}^{\infty}\frac{\alpha_n(\omega)\mathrm{i}}{\omega-\omega_n}e^{-\mathrm{i}\omega t}\,d\omega \\ &= \sum_{n=1}^{N} u_n(x)\alpha_n(\omega_n)e^{-\mathrm{i}\omega_n t}, \quad t>0. \end{aligned} \tag{3.40}$$

As an example, suppose that $p^{\mathrm{in}}(x,t)$ is a time-limited pulse of a plane wave with frequency $\omega_{\mathrm{in}} \in \mathbb{R}$ traveling in the x_1 direction. This is given by

$$p^{\mathrm{in}}(x,t) = e^{\mathrm{i}\omega_{\mathrm{in}}(x_1/v-t)}, \quad 0<t<1. \tag{3.41}$$

This has Fourier transform

$$u^{\mathrm{in}}(x,\omega) = 2e^{\frac{\mathrm{i}}{2}(\omega-\omega_{\mathrm{in}})}\operatorname{sinc}(\omega-\omega_{\mathrm{in}})e^{\mathrm{i}\omega_{\mathrm{in}}x_1/v}. \tag{3.42}$$

We can then compute $\alpha_1(\omega),\dots,\alpha_N(\omega)$ using (3.39). This is shown in Figure 3.6. We show, first, how the $L^2(D)$-norm of the solution to the scattering problem (3.3) varies as a function of ω_{in}. As is expected, the response is enhanced when ω_{in} is close to $\Re(\omega_n)$ for some $n = 1,\dots,N$. We also show how the weights $\alpha_1(\omega_1),\dots,\alpha_N(\omega_N)$ in (3.40) vary as a function of ω_{in}. Each constant is small except in a region of the associated resonant frequency when the corresponding eigenmode is excited most strongly. This demonstrates the mechanism through which the cochlea-inspired mechanism decomposes an incoming sound into its frequency components: by measuring the value of the acoustic pressure field on each resonator and then solving (3.39) to find the weight functions, we can identify when each of the resonant frequencies corresponds to a frequency component of the incoming signal. Increasing the number of resonators, we could achieve a frequency decomposition with resolution limited only by the widths of the peaks in Figure 3.6.

An addition consequence of the (formal) time-domain decomposition (3.40) is that we can understand how the system behaves in response to a transient signal. One of the most contentious issues in the development of cochlear models during the twentieth century was how to account for the fact that the cochlea demonstrates a "traveling wave" in response to a transient signal. It had been known since the mid-nineteenth century that the cochlea's function was based on graded elements being tuned to different audible frequencies, distributed along the length of the cochlea [86]. However, in the middle of the twentieth century famous experiments by Georg von Békésy demonstrated that when the cochlea is stimulated a wave travels from the base to the apex along the membrane upon which the receptor cells are mounted [161]. This discovery won him a

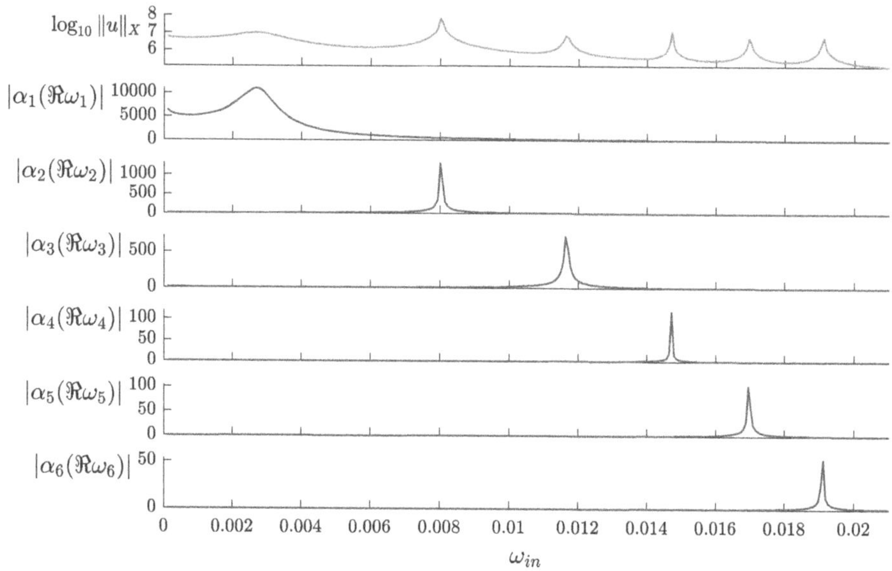

Figure 3.6: A system of six resonators filters an acoustic signal into the six subwavelength resonant frequencies. Here, a system of six circular resonators that increase in size is subjected to an incoming plane wave with frequency ω_{in}. The first plot shows how the norm of the solution $u(x, \omega)$ to (3.3) varies as a function of ω_{in}. We then show how each coefficient $\alpha_1(\omega_1), \ldots, \alpha_6(\omega_N)$ in (3.40) varies. The six resonant frequencies of this system are $\omega_1 = 0.002752 - 0.000538i$, $\omega_2 = 0.008026 - 0.000009i$, $\omega_3 = 0.011659 - 0.000048i$, $\omega_4 = 0.014703 - 0.000004i$, $\omega_5 = 0.016976 - 0.000009i$, $\omega_6 = 0.019096 - 0.000004i$.

Nobel Prize in 1961 and lead to the creation of a popular class of models based on each receptor cell being excited in series as the signal travels through the cochlea.

The main source of confusion related to this traveling wave is whether it is the main mechanism through which the signal propagates through the cochlea and how to resolve the fact that it travels at a speed much slower than the speed of sound in the background fluid. In Figure 3.7 we can see five snapshots of the acoustic pressure, plotted along the line through the resonators centers, when the system is excited by a transient signal. The existence of a wave traveling from the small high-frequency resonators at the start of the array to the larger low-frequency resonators at the far end is clear. This wave is the movement of the position of maximum acoustic pressure along the array of resonators. It is a consequence of the asymmetric eigenmodes (shown in Figures 3.3a–d) growing from rest at different rates.

The wave in Figure 3.7 shares a number of qualitative characteristics with the traveling wave observed (e. g., by Békésy) on the basilar membrane in the cochlea. For instance, the amplitude initially grows before quickly diminishing and the wave is seen to slow as it moves through the array [79, 161, 69]. It should also be noted that traveling waves have been observed in the cochlear fluid (as predicted for the resonator array considered here) as well as on the surface of the basilar membrane [133]. Further, this

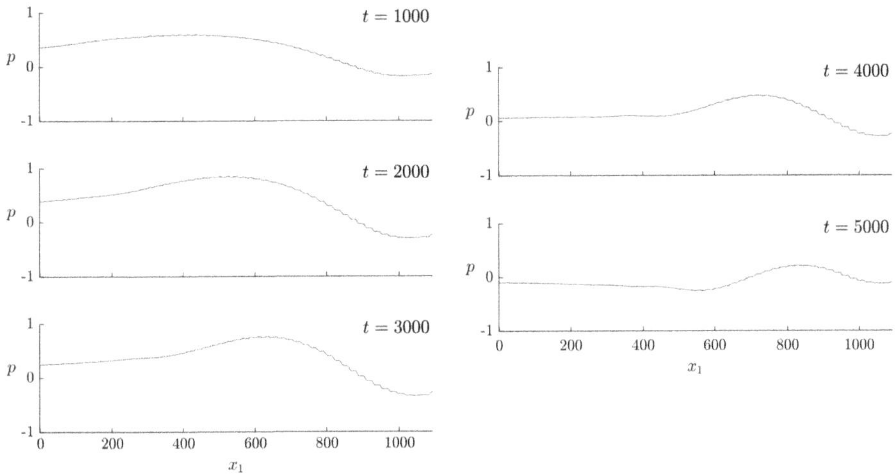

Figure 3.7: A graded metamaterial exhibits traveling wave behavior in response to a transient signal. Here, we show the evolution over time of the acoustic pressure scattered by 50 evenly spaced circular resonators with graded size (increasing from left to right). The acoustic pressure is initially zero then the resonant elements are excited at $t = 0$. We plot the cross-section of the pressure distribution along the straight line through the centers of the resonators.

wave is traveling much more slowly than the speed of sound in the surrounding fluid (which has been non-dimensionalized to unit for these numerics); the speed of sound in cochlear fluid is approximately 1500 ms^{-1} whereas the traveling wave is observed at speeds close to 10 ms^{-1} [34, 69]. The realization that coupled-resonator models reproduce this behavior has allowed some of the differences between previous cochlear models to be resolved [34].

3.7 Discussion

We have computed leading-order approximations to the resonant frequencies and associated eigenmodes for a system of coupled subwavelength acoustic resonators in two dimensions. We studied a resonator array that is graded in size and saw that it is able to perform so-called “rainbow trapping” (as introduced in Chapter 2) whereby different frequencies are reflected at different positions in the array, with a monotonic relationship between the frequency and the position. Further, we saw that this model has the ability to decompose incoming signals into these resonant modes, offering a platform for using these devices as a platform for an artificial hearing device.

4 3D metamaterial design

We now turn our attention to designing a graded metamaterial in three dimensions. While this is computationally slightly more challenging, the asymptotic analysis will be more straightforward than in the two-dimensional problems of the previous chapter.

4.1 Problem setting

The formulation we consider is the three-dimensional analogue of the two-dimensional problem from the previous chapter. In particular, we use the same notation for the material constants, and refer the reader to Section 3.1 for their definitions. The difference here is that the inclusions $D_1, \dots, D_N$ are subsets of $\mathbb{R}^3$ again having Hölder continuous boundaries. That is, each boundary ∂D_n is such that there is some $0 < s < 1$ so that $\partial D_n \in C^{1,s}$ (so it is locally the graph of a differentiable function whose derivatives are Hölder continuous with exponent s). The differential problem is

$$\begin{cases} (\Delta + k^2)u(x,\omega) = 0, & \text{in } \mathbb{R}^3 \setminus \overline{D}, \\ (\Delta + k_b^2)u(x,\omega) = 0, & \text{in } D, \\ u_+ - u_- = 0, & \text{on } \partial D, \\ \delta \frac{\partial u}{\partial \nu}\big|_+ - \frac{\partial u}{\partial \nu}\big|_- = 0, & \text{on } \partial D, \\ u^s := u - u^{\text{in}} \text{ satisfies the SRC}, & \text{as } |x| \to \infty. \end{cases} \tag{4.1}$$

In this three-dimensional case, the Sommerfeld radiation condition (SRC) is

$$\lim_{|x|\to\infty} |x| \left(\frac{\partial}{\partial |x|} - ik \right) u^s(x,\omega) = 0. \tag{4.2}$$

Once again, we will use boundary integral operators to study the solutions to (4.1). These are the obvious analogues of the two-dimensional operators introduced in Definition 3.1.1. We need to use the appropriate outgoing fundamental solution G^k, which is given by

$$G^k(x) = -\frac{1}{4\pi|x|} \exp(\mathrm{i}\omega|x|). \tag{4.3}$$

Then, we can define the corresponding single layer potential and Neumann–Poincaré operator in the same way as in two dimensions:

$$\mathcal{S}_D^k[\varphi](x) = \int_{\partial D} G^k(x-y)\varphi(y)\,\mathrm{d}\sigma(y), \quad x \in \partial D, \varphi \in L^2(\partial D), \tag{4.4}$$

$$\mathcal{K}_D^{k,*}[\varphi](x) = \int_{\partial D} \frac{\partial G^k(x-y)}{\partial \nu(x)} \varphi(y)\,\mathrm{d}\sigma(y), \quad x \in \partial D, \varphi \in L^2(\partial D). \tag{4.5}$$

https://doi.org/10.1515/9783110784961-004

These are related by the same jump relations as in the two-dimensional case:

$$\frac{\partial}{\partial \nu} \mathcal{S}_D^k[\varphi]\bigg|_{\pm} = \left(\pm \frac{1}{2} I + \mathcal{K}_D^{k,*} \right). \tag{4.6}$$

4.2 Asymptotic analysis

We will follow a similar asymptotic methodology to the one that was applied for the two-dimensional problem in Chapter 3. That is, we will let $\delta \to 0$ in (4.1) and seek subwavelength solutions for which $\omega \to 0$ also.

Since we are interested in low-frequency solutions, we will make use of the asymptotic expansions of the integral operators as $k \to 0$. The expansions in this case are slightly more convenient than in the two-dimensional setting. We have that

$$\mathcal{S}_D^k[\varphi] = \mathcal{S}_D[\varphi] + k\mathcal{S}_{D,1}[\varphi] + O(k^2), \tag{4.7}$$

where $\mathcal{S}_D := \mathcal{S}_D^0$ is given by (4.4) with $k = 0$ (i. e., the Laplace single-layer potential) and

$$\mathcal{S}_{D,1}[\varphi](x) := \frac{1}{4\pi \mathrm{i}} \int_{\partial D} \varphi(y) \, \mathrm{d}\sigma(y).$$

One crucial property to note is that $\mathcal{S}_D$ is invertible as a map $L^2(\partial D) \to H^1(\partial D)$. This is the main difference from the analysis presented in Chapter 3: the leading-order term in the expansion of the single layer potential is always invertible. This means we do not need to undertake the analysis presented in Lemmas 3.2.2 and 3.2.4 for the two-dimensional case.

As for the single layer potential, the Neumann–Poincaré operator (4.5) has a helpful asymptotic expansion at low frequencies. In particular, we have that as $k \to 0$,

$$\mathcal{K}_D^{k,*}[\varphi] = \mathcal{K}_D^*[\varphi] + k\mathcal{K}_{D,1}[\varphi] + k^2\mathcal{K}_{D,2}[\varphi] + k^3\mathcal{K}_{D,3}[\varphi] + O(k^4), \tag{4.8}$$

where $\mathcal{K}_D^* := \mathcal{K}_D^{0,*}$, $\mathcal{K}_{D,1} = 0$,

$$\mathcal{K}_{D,2}[\varphi](x) := \frac{1}{8\pi} \int_{\partial D} \frac{(x-y) \cdot \nu(x)}{|x-y|} \varphi(y) \, \mathrm{d}\sigma(y), \tag{4.9}$$

and

$$\mathcal{K}_{D,3}[\varphi](x) := \frac{\mathrm{i}}{12\pi} \int_{\partial D} (x-y) \cdot \nu(x) \varphi(y) \, \mathrm{d}\sigma(y). \tag{4.10}$$

Several of the operators in the expansion (4.8) can be simplified when integrated over all or part of the boundary ∂D. In particular, we have the following lemma, which was proved in [16, Lemma 2.1].

Lemma 4.2.1. *Given any $\varphi \in L^2(\partial D)$ and $i = 1,\dots,N$, it holds that*

$$\int_{\partial D_i} \left(-\frac{1}{2}I + \mathcal{K}_D^*\right)[\varphi]\, d\sigma = 0, \quad \int_{\partial D_i} \left(\frac{1}{2}I + \mathcal{K}_D^*\right)[\varphi]\, d\sigma = \int_{\partial D_i} \varphi\, d\sigma,$$

$$\int_{\partial D_i} \mathcal{K}_{D,2}[\varphi]\, d\sigma = -\int_{D_i} \mathcal{S}_D[\varphi]\, dx, \quad \textit{and} \quad \int_{\partial D_i} \mathcal{K}_{D,3}[\varphi]\, d\sigma = \frac{\mathrm{i}|D_i|}{4\pi} \int_{\partial D} \varphi\, d\sigma.$$

Given these results, we can proceed as in Section 3.2 to understand the subwavelength resonant modes of the system. In this case, the quantities that will emerge will be *capacitance coefficients*. Capacitance coefficients were first introduced by Maxwell to study the classical many-body problem in electrostatics of modeling the relationship between the distributions of potential and charge in a system of conductors [122]. More recently, it has been shown that the capacitance matrix model also captures the subwavelength resonant modes of a system of high-contrast resonators, as we will see here. It can also be thought of as a canonical model for a coupled system of three-dimensional resonators, which has long-range interactions decaying in proportion to the distance.

In order to introduce the notion of capacitance, we define the functions ψ_j, for $j = 1,\dots,N$, as

$$\psi_j := \mathcal{S}_D^{-1}[\chi_{\partial D_j}], \tag{4.11}$$

where $\mathcal{S}_D^{-1}$ is the inverse of $\mathcal{S}_D : L^2(\partial D) \to H^1(\partial D)$ and $\chi_A : \mathbb{R}^3 \to \{0,1\}$ is used to denote the characteristic function of a set $A \subset \mathbb{R}^3$. The capacitance matrix $C = (C_{ij})$ is defined, for $i,j = 1,\dots,N$, as

$$C_{ij} := -\int_{\partial D_i} \psi_j\, d\sigma. \tag{4.12}$$

It is known that C is symmetric [68]. In order to capture the behavior of an array of resonators with different sizes we need to introduce the generalized capacitance matrix $\mathcal{C} = (\mathcal{C}_{ij})$, given by

$$\mathcal{C}_{ij} := \frac{1}{|D_i|} C_{ij}, \tag{4.13}$$

which accounts for the differently sized resonators. There are many other variants of the generalized capacitance matrix, which can also be modified to account for different material parameters on the resonators, non-Hermitian systems and time-modulated systems, see, e. g., [10, 13]. It turns out that the eigenvalues of the generalized capacitance matrix determine the leading-order term in the asymptotic expansion of ω in terms of $\delta \to 0$. This concise characterization of the scattering problem will be the key that allows us to fine-tune the resonator array such that it replicates the action of the cochlea.

Many of the properties of the capacitance matrix C are well known [68]. It is an N-by-N positive-definite symmetric matrix with real-valued entries. While the weighting in the generalized capacitance matrix $\mathcal{C}$ hides some its underlying structure, it is the product of the symmetric matrix C with the diagonal matrix of inverse volumes (4.13). This means, in particular, that it has a basis of eigenvectors.

The relevance of the generalized capacitance matrix is that its eigenstates determine the resonant modes of the system, at leading order. This is analogous to the result that was derived in (3.24) for the two-dimensional version of the problem. However, in three dimensions the lack of logarithmic terms in the asymptotic expansion of $\mathcal{S}_D^k$ (4.7) means the leading-order equation can be reduced to an eigenvalue problem for $\mathcal{C}$ with eigenvalue proportional to ω^2. Before proving the main theorem, we derive a preliminary lemma that gives a decomposition of the solution to (4.1) in terms of a set of N functions. Let us define the functions S_n^ω, for $n = 1, \dots, N$, as

$$S_n^\omega(x) := \begin{cases} \mathcal{S}_D^k[\psi_n](x), & x \in \mathbb{R}^3 \setminus \overline{D}, \\ \mathcal{S}_D^{k_b}[\psi_n](x), & x \in D. \end{cases} \tag{4.14}$$

The significance of the capacitance matrix $\mathcal{C}$ will become apparent when we project the solution u onto the space spanned by these functions.

Lemma 4.2.2. *The solution to the scattering problem* (4.1) *can be written, for $x \in \mathbb{R}^3$ and incoming wave $u^{\mathrm{in}} = Ae^{\mathrm{i}kx_1}$, as*

$$u(x) - Ae^{\mathrm{i}kx_1} = \sum_{n=1}^N q_n S_n^\omega(x) - \mathcal{S}_D^k[\mathcal{S}_D^{-1}[Ae^{\mathrm{i}kx_1}]](x) + O(\omega),$$

for constants q_n which solve, up to an error of order $O(\delta\omega + \omega^3)$, the problem

$$(\omega^2 I_N - v_b^2 \delta\, \mathcal{C}) \begin{pmatrix} q_1 \\ \vdots \\ q_N \end{pmatrix} = v_b^2 \delta \begin{pmatrix} \frac{1}{|D_1|}\int_{\partial D_1} \mathcal{S}_D^{-1}[Ae^{\mathrm{i}kx_1}]\, \mathrm{d}\sigma \\ \vdots \\ \frac{1}{|D_N|}\int_{\partial D_N} \mathcal{S}_D^{-1}[Ae^{\mathrm{i}kx_1}]\, \mathrm{d}\sigma \end{pmatrix}, \tag{4.15}$$

where I_N is the $N \times N$ identity matrix.

Proof. The solutions can be represented as

$$u(x) = \begin{cases} Ae^{\mathrm{i}kx_1} + \mathcal{S}_D^k[\psi](x), & x \in \mathbb{R}^3 \setminus \overline{D}, \\ \mathcal{S}_D^{k_b}[\phi](x), & x \in D, \end{cases} \tag{4.16}$$

for some surface potentials $(\phi, \psi) \in L^2(\partial D) \times L^2(\partial D)$, which must be chosen so that u satisfies the transmission conditions across ∂D.

Using (3.10), we see that in order to satisfy the transmission conditions on ∂D the densities ϕ and ψ must satisfy

$$\mathcal{S}_D^{k_b}[\phi](x) - \mathcal{S}_D^{k}[\psi](x) = Ae^{\mathrm{i}kx_1}, \quad x \in \partial D,$$
$$\left(-\frac{1}{2}I + \mathcal{K}_D^{k_b,*}\right)[\phi](x) - \delta\left(\frac{1}{2}I + \mathcal{K}_D^{k,*}\right)[\psi](x) = \delta\frac{\partial}{\partial \nu}(Ae^{\mathrm{i}kx_1}), \quad x \in \partial D.$$

Using the asymptotic expansions (4.7) and (4.8), we can see that

$$\psi = \phi - \mathcal{S}_D^{-1}[Ae^{\mathrm{i}kx_1}] + O(\omega),$$

and, further, that up to an error of order $O(\delta\omega + \omega^3)$,

$$\left(-\frac{1}{2}I + \mathcal{K}_D^* + \frac{\omega^2}{v_b^2}\mathcal{K}_{D,2} - \delta\left(\frac{1}{2}I + \mathcal{K}_D^*\right)\right)[\phi] = -\delta\left(\frac{1}{2}I + \mathcal{K}_D^*\right)\mathcal{S}_D^{-1}[Ae^{\mathrm{i}kx_1}]. \tag{4.17}$$

At leading order, (4.17) says that $(-\frac{1}{2}I + \mathcal{K}_D^*)[\phi] = 0$ so, in light of the fact that $\{\psi_1, \dots, \psi_N\}$ forms a basis for $\ker(-\frac{1}{2}I + \mathcal{K}_D^*)$, the solution can be written as

$$\phi = \sum_{n=1}^{N} q_n\psi_n + O(\omega^2 + \delta), \tag{4.18}$$

for constants $q_1, \dots, q_N = O(1)$. Then, integrating (4.17) over ∂D_i, for $1 \le i \le N$, and using the properties (4.2.1) gives us that

$$-\omega^2 \int_{D_i} \mathcal{S}_D[\phi]\, \mathrm{d}x - v_b^2\delta \int_{\partial D_i} \phi\, \mathrm{d}\sigma = -v_b^2\delta \int_{\partial D_i} \mathcal{S}_D^{-1}[Ae^{\mathrm{i}kx_1}]\, \mathrm{d}\sigma + O(\delta\omega + \omega^3).$$

Substituting the ansatz (4.18) gives the desired result. □

Recall that the resonant modes of the system are nontrivial solutions that exist in the case that $u^{\mathrm{in}} = 0$. Thus, from Lemma 4.2.2, we can see that resonance occurs when $\omega^2/v_b^2\delta$ is an eigenvalue of C, at leading order. We can continue this argument to higher orders and obtain the following theorem, which characterizes the subwavelength resonant frequencies of the system.

Theorem 4.2.1. *As $\delta \to 0$, the subwavelength resonant frequencies satisfy the asymptotic formula*

$$\omega_n^{\pm} = \pm\sqrt{v_b^2\lambda_n\delta} - \mathrm{i}\tau_n\delta + O(\delta^{3/2}),$$

for $n = 1, \dots, N$, where λ_n are the eigenvalues of the generalized capacitance matrix C and τ_n are nonnegative real numbers given by

$$\tau_n = \frac{v_b^2}{8\pi v}\frac{\underline{v}_n \cdot CJC\underline{v}_n}{\|\underline{v}_n\|_D^2},$$

where C is the capacitance matrix, J is the $N \times N$ matrix of ones, $\underline{v}_n$ is the eigenvector associated to λ_n and we use the norm $\|x\|_D := (\sum_{i=1}^N |D_i| x_i^2)^{1/2}$.

Proof. If $u^{\text{in}} = 0$, we find from Lemma 4.2.2 that there is a nonzero solution $q_1, \dots, q_N$ to the eigenvalue problem (4.15) when $\omega^2/v_b^2\delta$ is an eigenvalue of C, at leading order.

To find the imaginary part, we adopt the ansatz

$$\omega_n^{\pm} = \pm\sqrt{v_b^2 \lambda_n \delta} - \mathrm{i}\tau_n \delta + O(\delta^{3/2}), \tag{4.19}$$

where λ_n is an eigenvalue of C and τ_n is a real number. Using the expansions (4.7) and (4.8) with the representation (4.16), we have that

$$\psi = \phi + \frac{k_b - k}{4\pi \mathrm{i}}\left(\sum_{n=1}^{N} \psi_n\right) \int_{\partial D} \phi \, \mathrm{d}\sigma + O(\omega^2),$$

and hence that

$$\begin{aligned}\left(-\frac{1}{2}I + \mathcal{K}_D^* + k_b^2 \mathcal{K}_{D,2} + k_b^3 \mathcal{K}_{D,3} - \delta\left(\frac{1}{2}I + \mathcal{K}_D^*\right)\right)[\phi] \\ - \frac{\delta(k_b - k)}{4\pi \mathrm{i}}\left(\sum_{n=1}^{N} \psi_n\right) \int_{\partial D} \phi \, \mathrm{d}\sigma = O(\delta\omega^2 + \omega^4).\end{aligned} \tag{4.20}$$

We then substitute the decomposition (4.18) and integrate over ∂D_i, for $i = 1, \dots, N$, to find that, up to an error of order $O(\delta\omega^2 + \omega^4)$, it holds that

$$\left(-\frac{\omega^2}{v_b^2} - \frac{\omega^3 \mathrm{i}}{4\pi v_b^3} JC + \delta C + \frac{\delta\omega \mathrm{i}}{4\pi}\left(\frac{1}{v_b} - \frac{1}{v}\right) CJC\right)\underline{q} = 0,$$

where J is the $N \times N$ matrix of ones (i. e., $J_{ij} = 1$ for all $i, j = 1, \dots, N$). Then, using the ansatz (4.19) for $\omega_n^{\pm}$, we see that if $\underline{v}_n$ is an eigenvector corresponding to λ_n, then it holds that

$$\tau_n = \frac{v_b^2}{8\pi v} \frac{\underline{v}_n \cdot CJC\underline{v}_n}{\|\underline{v}_n\|_D^2}, \tag{4.21}$$

where we use the norm $\|x\|_D := (\sum_{i=1}^N |D_i| x_i^2)^{1/2}$ for $x \in \mathbb{R}^N$. Since C is symmetric, we can see that $\tau_n \geq 0$. □

Given the importance of Theorem 4.2.1 for the analysis that will follow, we pause to make some remarks about it. Firstly, we observe that the fact that $\omega \propto \sqrt{\delta}$ is a direct consequence of the structure of the asymptotic expansion of $\mathcal{K}_D^{k,*}$. In particular, the fact that $\mathcal{K}_{D,1} = 0$ implies that $\mathcal{K}_D^{k,*} = \mathcal{K}_D^{0,*} + O(k^2)$ so the perturbation to the leading-order solution occurs at order $O(k^2)$. Secondly, the fact that the leading-order imaginary part is nonpositive in Theorem 4.2.1 (equivalently, that $\tau_n \geq 0$) follows from C being symmetric, since then $\underline{v}_n \cdot CJC\underline{v}_n = (C\underline{v}_n) \cdot J(C\underline{v}_n)$ and J is positive semidefinite. Physically, this corresponds to the fact that energy is lost to the far field (with the magnitude of this negative imaginary part describing the rate of attenuation).

4.3 Numerical methods

An analogue of the multipole expansion approach used in Section 3.3 can be generalized to this setting, provided we choose the right basis. Since the analysis is similar for both the case of finitely many resonators and an infinite periodic system of resonators, we will derive the multipole expansion approximation of $\mathcal{S}_D^k$ and $\mathcal{S}_D^{\alpha,k}$ in three dimensions. This will allow us to compute both the resonance of an array of finitely many resonators and also the spectral bands for a periodic array, as depicted in Figures 2.2 and 2.3. The operator $\mathcal{S}_D^{\alpha,k}$ is the analogous single layer potential for the periodic problem with quasiperiodic boundary conditions, i. e., the system of equations (2.4). It is defined analogously to the standard single layer potential, with a Green's function that is modified by taking the Floquet transform (2.1) of the standard Green's function. For the simplest example of a three-dimensional problem which is periodic in one dimension, we can define the unit cell Y as $Y := Y_0 \times \mathbb{R}^2$. We define the quasiperiodic Green's function $G^{\alpha,k}(x)$ as the Floquet transform of the outgoing Helmholtz Green's function $G^k(x)$ in the first dimension of x, i. e.,

$$G^{\alpha,k}(x) := -\sum_{m\in\mathbb{Z}} \frac{e^{ik|x-(Lm,0,0)|}}{4\pi|x-(Lm,0,0)|} e^{i\alpha Lm}. \tag{4.22}$$

Let D be as in the previous layer potential definitions, but assume additionally $D \Subset Y$. We define the quasiperiodic single layer potential $\mathcal{S}_D^{\alpha,k}$ by

$$\mathcal{S}_D^{\alpha,k}[\phi](x) := \int_{\partial D} G^{\alpha,k}(x-y)\phi(y)\, \mathrm{d}\sigma(y), \quad x \in \mathbb{R}^3. \tag{4.23}$$

It is known that $\mathcal{S}_D^{\alpha,0} : L^2(\partial D) \to H^1(\partial D)$ is invertible if $\alpha \neq 0$ [14]. It satisfies the jump relations

$$\mathcal{S}_D^{\alpha,k}[\phi]|_+ = \mathcal{S}_D^{\alpha,k}[\phi]|_-, \tag{4.24}$$

and

$$\frac{\partial}{\partial \nu}\mathcal{S}_D^{\alpha,k}[\phi]\Big|_{\pm} = \left(\pm\frac{1}{2}I + (\mathcal{K}_D^{-\alpha,k})^*\right)[\phi] \quad \text{on } \partial D, \tag{4.25}$$

where $(\mathcal{K}_D^{-\alpha,k})^*$ is the quasiperiodic Neumann–Poincaré operator, given by

$$(\mathcal{K}_D^{-\alpha,k})^* [\phi](x) := \int_{\partial D} \frac{\partial}{\partial \nu_x} G^{\alpha,k}(x-y)\phi(y)\, d\sigma(y). \tag{4.26}$$

The quasiperiodic single layer potential satisfies a low-frequency asymptotic expansion given by

$$\mathcal{S}_\Omega^{\alpha,k} = \mathcal{S}_\Omega^{\alpha,0} + O(k^2). \tag{4.27}$$

The method is a generalization of the method in two dimensions given in Section 3.3. The overarching principle is that when working on spherical domains, the action of the single layer potential on spherical basis functions has an explicit, analytic representation. In three dimensions, the integral operator we need to discretize is slightly simpler. Instead of having to treat the matrix-valued operator $\mathcal{A}$ as in (3.12), we can see that the invertibility of the single layer potential allows us to discard the first row of this system and deal immediately with the problem

$$\mathcal{A}(\omega, \delta)[\psi] = 0, \tag{4.28}$$

where

$$\mathcal{A}(\omega, \delta) := -\lambda I + \mathcal{K}_D^{\omega,*}, \quad \lambda := \frac{1+\delta}{2(1-\delta)}. \tag{4.29}$$

The analogous operator for the periodic problem (2.4) is

$$\mathcal{A}^\alpha(\omega, \delta)[\psi^\alpha] = 0, \tag{4.30}$$

where

$$\mathcal{A}^\alpha(\omega, \delta) := -\lambda I + (\mathcal{K}_D^{-\alpha,\omega})^*, \quad \lambda := \frac{1+\delta}{2(1-\delta)}. \tag{4.31}$$

The goal is to discretize equations (4.28) and (4.30). Using the jump relations (4.6) and (4.25), observe that the operators $\mathcal{A}$ and $\mathcal{A}^\alpha$ can be written as

$$\mathcal{A}(k, \delta) = \frac{\partial}{\partial \nu} \mathcal{S}_D^k \Big|_- - \delta \frac{\partial}{\partial \nu} \mathcal{S}_D^k \Big|_+$$

and

$$\mathcal{A}^\alpha(k, \delta) = \frac{\partial}{\partial \nu} \mathcal{S}_D^{\alpha,k} \Big|_- - \delta \frac{\partial}{\partial \nu} \mathcal{S}_D^{\alpha,k} \Big|_+,$$

so it is enough to find a discretized representation of the single layer potentials $\mathcal{S}_D^k$ and $\mathcal{S}_D^{\alpha,k}$.

For a radially symmetric Helmholtz equation in three dimensions, it is well known that the spherical waves $j_l(kr)Y_l^m(\theta,\phi)$ and $h_l^{(1)}(kr)Y_l^m(\theta,\phi)$ give a basis of solutions in the spherical polar coordinates (r, θ, ϕ). Here $Y_l^m(\theta, \phi), l \in \mathbb{N}, m = -l, \ldots, l$, are the spherical harmonics and $j_n(kr)$, $h_n^{(1)}(kr)$ are the spherical Bessel and Hankel functions of the first kind, respectively, defined by

$$j_l(x) = \sqrt{\frac{\pi}{2x}} J_{l+\frac{1}{2}}(x), \quad h_l^{(1)}(x) = \sqrt{\frac{\pi}{2x}} H_{l+\frac{1}{2}}^{(1)}(x),$$

where J_n and $H_n^{(1)}$ are the ordinary Bessel and Hankel functions of the first kind [1].

We begin by deriving the multipole expansion of the single layer potential $\mathcal{S}_D^k$. The spherical harmonics Y_l^m form a basis of $L^2(\partial D)$ and we seek the expansion of $\mathcal{S}_D^k$ in this basis. Define $u := \mathcal{S}_D^k[Y_l^m]$, which is the solution to

$$\begin{cases} \Delta u + k^2 u = 0 & \text{in } \mathbb{R}^3 \setminus \overline{D}, \\ \Delta u + k^2 u = 0 & \text{in } D, \\ u|_+ = u|_- & \text{on } \partial D, \\ \frac{\partial u}{\partial \nu}\big|_+ - \frac{\partial u}{\partial \nu}\big|_- = Y_l^m & \text{on } \partial D, \\ |x|\left(\frac{\partial}{\partial |x|} - \mathrm{i}k\right)u \to 0 & \text{as } |x| \to \infty. \end{cases} \tag{4.32}$$

The above equation can be easily solved by the separation of variables technique in polar coordinates. It gives

$$\mathcal{S}_D^k[Y_l^m](r, \theta, \phi) = \begin{cases} c j_l(kR) h_l^{(1)}(kr) Y_l^m(\theta, \phi), & |r| > R, \\ c h_l^{(1)}(kR) j_l(kr) Y_l^m(\theta, \phi), & |r| \leq R, \end{cases} \tag{4.33}$$

where $c = -\mathrm{i}kR^2$.

In order to handle problems posed on disjoint domains, with multiple resonators, we will need an addition theorem relating spherical waves centered around a translated origin to spherical waves around the original origin. Suppose we have a point with coordinates $x = (r, \theta, \phi)$ in the original system and $x' = (r', \theta', \phi')$ in the translated system, with the coordinate vectors related by $x = x' + b$ for $b = (r_b, \theta_b, \phi_b)$. Moreover, we assume $r' < r_b$. Then, the addition theorem reads [76]

$$h_l^{(1)}(kr) Y_l^m(\theta, \phi) = \sum_{l' \in \mathbb{N}, |m'| \leq l'} A_{l'm'}^{lm} j_{l'}(kr') Y_{l'}^{m'}(\theta', \phi'), \tag{4.34}$$

where the coefficients $A_{l'm'}^{lm}$ are given by

$$A_{l'm'}^{lm} = \sum_{\lambda \in \mathbb{N}, |\mu| \leq \lambda} C(l, m, l', m', \lambda, \mu) h_\lambda^{(1)}(kr_b) Y_\lambda^\mu(\theta_b, \phi_b).$$

Here, the coefficients $C(l, m, l', m', \lambda, \mu)$ are in turn given by

$$C(l,m,l',m',\lambda,\mu) = i^{l'-l+\lambda}(-1)^m\sqrt{4\pi(2l+1)(2l'+1)(2\lambda+1)} \times \begin{pmatrix} l & l' & \lambda \\ 0 & 0 & 0 \end{pmatrix}\begin{pmatrix} l & l' & \lambda \\ -m & m' & \mu \end{pmatrix}, \tag{4.35}$$

where the quantities of the form

$$\begin{pmatrix} j_1 & j_2 & j_3 \\ m_1 & m_2 & m_3 \end{pmatrix},$$

are Wigner 3j symbols: coefficients originally developed in quantum-mechanical settings to handle sums of angular momenta [166]. To simplify these expressions slightly, we assume that the original coordinate system is aligned such that the vector $b = (r_b, \theta_b, \phi_b)$ points along the positive z-axis, i. e., $\theta_b = 0$. In this case, we have that

$$Y_\lambda^\mu(\theta_b, \phi_b) = \begin{cases} 0, & \mu \neq 0, \\ \sqrt{\frac{2\lambda+1}{4\pi}}, & \mu = 0. \end{cases}$$

Substituting this into the expression for $A_{l'm'}^{lm}$ gives

$$A_{l'm'}^{lm} = \sum_{\lambda\in\mathbb{N}} \sqrt{\frac{2\lambda+1}{4\pi}} C(l,m,l',m',\lambda,0) j_\lambda(kr_b).$$

Now, we compute the quasiperiodic single layer potential $\mathcal{S}_D^{\alpha,k}[Y_l^m]$ in the case when D consists of a single resonator that repeats periodically according to the given lattice. Since

$$G^{\alpha,k}(x,y) = \sum_{n\in\mathbb{Z}} G^k(|x-y-(nL,0,0)|)e^{\mathrm{i}n\alpha L},$$

we have

$$\begin{aligned} \mathcal{S}_D^{\alpha,k}[Y_l^m](x) &= \mathcal{S}_D^k[Y_l^m](x) + \sum_{n\in\mathbb{Z},n\neq 0} \mathcal{S}_{D+n}^k[Y_l^m]e^{\mathrm{i}n\alpha L} \\ &= \mathcal{S}_D^k[Y_l^m](x) + cj_n(kR) \sum_{n\in\mathbb{Z},n\neq 0} h_l^{(1)}(kr_n')Y_l^m(\theta_n', \phi_n')e^{\mathrm{i}n\alpha}. \end{aligned}$$

Here, $D+n$ means a translation of the disk D by $(nL, 0, 0)$ and $(r_n', \theta_n', \phi_n')$ are the spherical coordinates with respect to the center of $D+n$.

Using the addition theorem (4.34), we have

$$\begin{aligned} \mathcal{S}_D^{\alpha,k}[Y_l^m](x) &= \mathcal{S}_D^k[Y_l^m](x) \\ &\quad + cj_l(kR) \sum_{l'\in\mathbb{N},|m'|\leq l'} \left[\sum_{\lambda\in\mathbb{N},|\mu|\leq\lambda} C(l,m,l',m',\lambda,\mu)Q_\lambda^\mu\right] j_{l'}(kr)Y_{l'}^{m'}(\theta,\phi) \end{aligned}$$

$$:= \mathcal{S}_D^k[Y_l^m](x) + cj_l(kR) \sum_{l' \in \mathbb{N}, |m'| \leq l'} B_{l'm'}^{lm} j_{l'}(kr) Y_{l'}^{m'}(\theta, \phi),$$

where Q_λ^μ is the one-dimensional lattice sum in three dimensions, defined by

$$Q_\lambda^\mu = \sum_{n \in \mathbb{Z}, n \neq 0} h_\lambda^{(1)}(knL) Y_\lambda^\mu(\theta_n, \phi_n) e^{\mathrm{i}n\alpha L}.$$

These lattice sums typically converge very slowly, making accurate numerical implementation difficult and expensive. However, alternative representations have been developed which overcome these difficulties. For example, an efficient method for computing this lattice sum can be found in [113].

We are now ready to compute the periodic single layer potential $\mathcal{S}_D^{\alpha,k}$ in the case when $D \Subset Y$ consists of two resonators, centered at $(-x_1, 0, 0)$ and $(x_1, 0, 0)$, respectively, as a demonstrative example. The generalization to systems of many resonators is a natural continuation of this, but is omitted for brevity. By identifying $L^2(\partial D) = L^2(\partial D_1) \times L^2(\partial D_2)$, we have

$$\mathcal{S}_D^{\alpha,k} = \begin{pmatrix} \mathcal{S}_{D_1}^{\alpha,k} & \mathcal{S}_{D_2}^{\alpha,k}|_{\partial D_1} \\ \mathcal{S}_{D_1}^{\alpha,k}|_{\partial D_2} & \mathcal{S}_{D_2}^{\alpha,k} \end{pmatrix}.$$

Here, the operator $\mathcal{S}_{D_i}^{\alpha,k}|_{\partial D_j} : L^2(\partial D_i) \to L^2(\partial D_j), i, j = 1, 2$ is the evaluation of $\mathcal{S}_{D_i}^{\alpha,k}$ on ∂D_j. To compute the multipole expansion of $\mathcal{S}_{D_1}^{\alpha,k}|_{\partial D_2}$, we again use the addition theorem. We have

$$\begin{aligned}
&\mathcal{S}_{D_1}^{\alpha,k}|_{\partial D_2}[Y_l^m](x') \\
&\quad = cj_l(kR) h_l^{(1)}(kr') Y_l^m(\theta', \phi') + cj_l(kR) \sum_{\substack{l' \in \mathbb{N} \\ |m'| \leq l'}} B_{l'm'}^{lm} j_{l'}(kr') Y_{l'}^{m'}(\theta', \phi') \\
&\quad = cj_l(kR) \sum_{\substack{l'' \in \mathbb{N} \\ |m''| \leq l''}} \left[\sum_{\substack{\lambda \in \mathbb{N} \\ |\mu| \leq \lambda}} C(l, m, l'', m'', \lambda, \mu) h_\lambda^{(1)}(kd) Y_\lambda^\mu(\theta_d, \phi_d) \right] j_{l''}(kr) Y_{l''}^{m''}(\theta, \phi) \\
&\qquad + cj_l(kR) \sum_{\substack{l'' \in \mathbb{N} \\ |m''| \leq l''}} \left[\sum_{\substack{l' \in \mathbb{N}, |m'| \leq l' \\ \lambda \in \mathbb{N}, |\mu| \leq \lambda}} B_{l'm'}^{lm} C(l', m', l'', m'', \lambda, \mu) j_\lambda(kd) Y_\lambda^\mu(\theta_d, \phi_d) \right] \\
&\qquad \times j_{l''}(kr) Y_{l''}^{m''}(\theta, \phi).
\end{aligned}$$

In order to simulate the array of a finite number of resonators, we must now perform similar computations for the operator $\mathcal{S}_D^k$ in the case when D consists of N resonators. We assume the resonators to be arranged collinearly along the x_1-axis, since this is sufficient for the examples we will consider here and it simplifies the form of the coefficients, thanks to the addition theorem (4.34). By identifying $L^2(\partial D) = L^2(\partial D_1)\times\cdots\times L^2(\partial D_N)$, we have

$$\mathcal{S}_D^k = \begin{pmatrix} \mathcal{S}_{D_1}^k & \mathcal{S}_{D_2}^k|_{\partial D_1} & \cdots & \mathcal{S}_{D_N}^k|_{\partial D_1} \\ \mathcal{S}_{D_1}^k|_{\partial D_2} & \mathcal{S}_{D_2}^k & \cdots & \mathcal{S}_{D_N}^k|_{\partial D_2} \\ \vdots & \vdots & \ddots & \vdots \\ \mathcal{S}_{D_1}^k|_{\partial D_N} & \mathcal{S}_{D_2}^k|_{\partial D_N} & \cdots & \mathcal{S}_{D_N}^k \end{pmatrix}, \tag{4.36}$$

where, as in the quasiperiodic case, $\mathcal{S}_{D_i}^k|_{\partial D_j} : L^2(\partial D_i) \to L^2(\partial D_j)$ is the evaluation of $\mathcal{S}_{D_i}^k$ on ∂D_j. This relies on the addition theorem once again. The diagonal terms are easily evaluated using (4.33) directly. Away from the diagonals, the addition theorem (4.34) gives that

$$\mathcal{S}_{D_j}^k|_{\partial D_i}[Y_l^m](x') = ch_l^{(1)}(kR) \sum_{l'\in\mathbb{N},|m'|\leq l'} A_{l'm'}^{lm} j_{l'}(kr') Y_{l'}^{m'}(\theta',\phi').$$

Once we have found a matrix representation for the operator we are interested in, it remains to use numerical root finding techniques to find the values of k for which the operator has a nontrivial kernel. As for the two-dimensional case, Muller's method is a good option, since it does not require us to compute any derivatives. The details of this were presented in Section 4.3.

4.4 Discussion

We now have both an asymptotic and a numerical method for modeling an array of three-dimensional high-contrast resonators. It is valuable to compare the virtues of the two methods. A comparison is displayed in Figure 4.1, for example. Here, we simulate the subwavelength resonant modes of the toy example of ten spherical resonators with unit radius arranged in a line. The plot shows both the numerical values, obtained using the multipole expansion method outlined in Section 4.3, and the asymptotic values, computed in terms of the eigenvalues of the generalized capacitance matrix, as set out in Theorem 4.2.1. In this case, we take $\delta = 1/5000$ and see good agreement between the two methods. The key difference between the two methods is the computational expense. In this case, running on a laptop, the computations using the full multipole method took 41 seconds while the approximations from the capacitance matrix took just 0.02 seconds. The reason for this is the need to use a numerical root finding method for the numerical approach. By removing the need to do so, the asymptotic method can reduce the computational expense by several orders of magnitude.

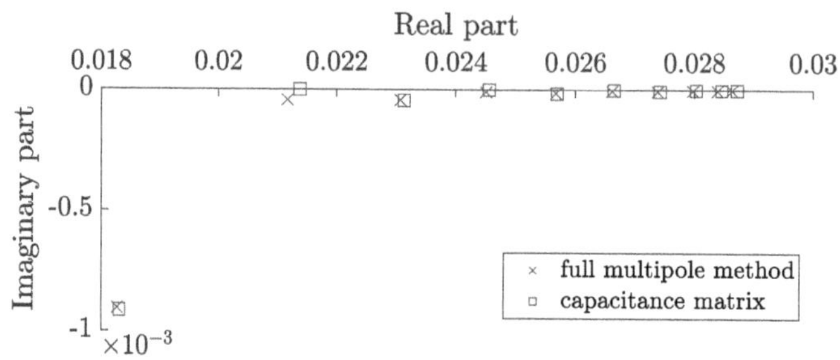

Figure 4.1: The subwavelength resonant frequencies of a system of ten spherical resonators. We compare the values computed using the multipole expansion method to discretize the full boundary integral equation and the values computed using the capacitance matrix. The computations using the full multipole method took 41 seconds while the approximations from the capacitance matrix took just 0.02 seconds, on the same computer. Each resonator has unit radius and we use $\delta = 1/5000$.

5 Implications for signal processing

In the previous chapters, we have developed methods for designing metasurfaces and metamaterials efficiently, based on asymptotic methods. As a result, we are able to realize graded structures that replicate the spatial frequency separation performed by the cochlea. A thought-provoking observation is that the underlying objective of this work, which is to develop systems that mimic the function of human auditory processing, is also shared by other communities. While we have considered physical wave-filtering devices so far in this work, many researchers and engineers have pursued a similar goal using computer algorithms. This field is typically known as *signal processing* and it is interesting to consider if we can form any precise links between cochlea-inspired metamaterials and bio-inspired signal processing algorithms. This comparison will be facilitated by the fact that we have analytic formulas for the scattering of an acoustic wave by a given high-contrast metamaterial.

5.1 Modal decompositions of signals

In the previous chapters, we only considered the scattering of time-harmonic waves, thereby allowing us to work solely in the frequency domain. Consider now the scattering of a more general signal $s : [0, T] \to \mathbb{R}$, whose frequency support is wider than a single frequency and whose Fourier transform exists. Again, we assume that the wave is incident parallel to the x_1-axis. Consider the Fourier transform of the incoming pressure wave, given for $\omega \in \mathbb{C}$, $x \in \mathbb{R}^3$ by

$$\begin{aligned} u^{\text{in}}(x, \omega) &= \int_{-\infty}^{\infty} s(x_1/v - t)e^{\mathrm{i}\omega t}\,\mathrm{d}t \\ &= e^{\mathrm{i}\omega x_1/v}\hat{s}(\omega) = \hat{s}(\omega) + O(\omega), \end{aligned} \tag{5.1}$$

as $\omega \to 0$, where $\hat{s}(\omega) := \int_{-\infty}^{\infty} s(u)e^{-\mathrm{i}\omega u}\,\mathrm{d}u$. The resulting pressure field satisfies the Helmholtz equation along with the transmission and radiation conditions.

The starting point for our analysis is Lemma 4.2.2, which describes how a wave scattered by an array of N high-contrast resonators can be decomposed into a space with dimension N, at leading order. However, it is more illustrative to rephrase Lemma 4.2.2 in terms of basis functions that are associated with the resonant frequencies. Let $V = (v_{i,j})$ be the matrix whose columns are the eigenvectors of the generalized capacitance matrix $\mathcal{C}$, as defined in (4.13). Recalling the fact that $\mathcal{C}$ has a basis of eigenvectors, we know that V is invertible. Then, we define the functions

$$u_n(x) = \sum_{i=1}^{N} v_{i,n}\,\mathcal{S}_D[\psi_i](x), \tag{5.2}$$

https://doi.org/10.1515/9783110784961-005

for $n = 1,\dots,N$, where the functions ψ_j were defined in (4.11). We will seek a modal decomposition in terms of these functions. We expect the coefficients to depend on the proximity of the frequency ω to the system's resonant frequencies $\omega_n^\pm$. With this in mind, we obtain the following lemma by diagonalizing C (with the change of basis matrix V) and solving the resulting system. The result has been simplified further by noticing that $\omega^2 - v_b^2\delta\lambda_n = (\omega - \omega_n^+)(\omega - \omega_n^-) + O(\omega^3)$ and that $e^{ikx_1} = 1 + ikx_1 + \cdots = 1 + O(\omega)$.

Lemma 5.1.1. *If $\omega = O(\sqrt{\delta})$, the solution to the scattering problem* (4.1) *with incoming wave $u^{\text{in}} = Ae^{ikx_1}$ can be written, for $x \in \mathbb{R}^3$, as*

$$u(x) - Ae^{ikx_1} = \sum_{n=1}^{N} a_n u_n(x) - \mathcal{S}_D[\mathcal{S}_D^{-1}[Ae^{ikx_1}]](x) + O(\omega),$$

for constants which satisfy, up to an error of order $O(\omega^3)$, the equations

$$a_n(\omega - \omega_n^+)(\omega - \omega_n^-) = -Av_n\mathbb{R}(\omega_n^+)^2,$$

where $v_n = \sum_{j=1}^{N}[V^{-1}]_{n,j}$, i. e., the sum of the nth row of V^{-1}.

Working in the frequency domain, the scattered acoustic pressure field u in response to the Fourier transformed signal $\hat{s}$ can be decomposed in the spirit of Lemma 5.1.1. We propose the ansatz that, for $x \in \partial D$, the solution to the scattering problem is given by the modal decomposition

$$u(x,\omega) = \sum_{n=1}^{N} \frac{-\hat{s}(\omega)v_n\mathbb{R}(\omega_n^+)^2}{(\omega - \omega_n^+)(\omega - \omega_n^-)} u_n(x) + r(x,\omega), \tag{5.3}$$

for some remainder r. We are interested in signals whose energy is mostly concentrated within the subwavelength regime. In particular, we will consider signals that are *subwavelength* in the sense that

$$\sup_{x\in\mathbb{R}^3} \int_{-\infty}^{\infty} |r(x,\omega)|\, d\omega = O(\delta). \tag{5.4}$$

This subwavelength condition (5.4) is a strong assumption and is difficult to interpret physically. However, for the purpose of seeking to inform signal processing algorithms, which is our aim here, it is a suitable assumption.

Now, we wish to apply the inverse Fourier transform to (5.3) to obtain a time-domain decomposition of the scattered field. The condition (5.4) guarantees that the remainder term is not significant. Meanwhile, the contributions from each term in the expansion can be found through complex integration.

Theorem 5.1.1. *For $\delta > 0$ and a signal s which is subwavelength in the sense of the condition* (5.4), *it holds that the scattered pressure field $p(x,t)$ satisfies, for $x \in \partial D$, $t \in \mathbb{R}$,*

$$p(x,t) = \sum_{n=1}^{N} a_n[s](t)u_n(x) + O(\delta),$$

where the coefficients are given by $a_n[s](t) = (s * h_n)(t)$ *for kernels defined as*

$$h_n(t) = \begin{cases} 0, & t < 0, \\ c_n e^{\Im(\omega_n^+)t} \sin(\Re(\omega_n^+)t), & t \geq 0, \end{cases} \tag{5.5}$$

for $c_n = v_n \Re(\omega_n^+)$.

Proof. Applying the inverse Fourier transform to the modal expansion (5.3) under the assumption (5.4) yields

$$p(x,t) = \sum_{n=1}^{N} a_n[s](t)u_n(x) + O(\delta),$$

where, for $n = 1, \dots, N$, the coefficients are given by

$$a_n[s](t) = \frac{1}{2\pi} \int_{-\infty}^{\infty} \frac{-\hat{s}(\omega) v_n \Re(\omega_n^+)^2}{(\omega - \omega_n^+)(\omega - \omega_n^-)} e^{-i\omega t} \, d\omega = (s * h_n)(t).$$

Here, $*$ denotes convolution and the kernels h_n are defined for $n = 1, \dots, N$ by

$$h_n(t) = \frac{1}{2\pi} \int_{-\infty}^{\infty} \frac{-v_n \Re(\omega_n^+)^2}{(\omega - \omega_n^+)(\omega - \omega_n^-)} e^{-i\omega t} \, d\omega. \tag{5.6}$$

We can use complex integration to evaluate the integral in (5.6). For $R > 0$, let $\Gamma_R^\pm$ be the semicircular arc of radius R in the upper $(+)$ and lower $(-)$ half-planes and let $\Gamma^\pm$ be the closed contour $\Gamma^\pm = \Gamma_R^\pm \cup [-R, R]$. Then, we have that

$$h_n(t) = \frac{1}{2\pi} \oint_{\Gamma^\pm} \frac{-v_n \Re(\omega_n^+)^2}{(\omega - \omega_n^+)(\omega - \omega_n^-)} e^{-i\omega t} \, d\omega - \frac{1}{2\pi} \int_{\Gamma_R^\pm} \frac{-v_n \Re(\omega_n^+)^2}{(\omega - \omega_n^+)(\omega - \omega_n^-)} e^{-i\omega t} \, d\omega.$$

The integral around $\Gamma^\pm$ is easy to evaluate using the residue theorem, since it has simple poles at $\omega_n^\pm$. We will make the choice of $+$ or $-$ so that the integral along $\Gamma_R^\pm$ converges to zero as $R \to \infty$. For large R, we have a bound of the form

$$\left| \int_{\Gamma_R^\pm} \frac{-v_n \Re(\omega_n^+)^2}{(\omega - \omega_n^+)(\omega - \omega_n^-)} e^{-i\omega t} \, d\omega \right| \leq C_n R^{-1} \sup_{\omega \in \Gamma_R^\pm} e^{\Im(\omega)t}, \tag{5.7}$$

for a positive constant C_n.

Suppose first that $t < 0$. Then we choose to integrate over Γ_R^+ in the upper complex plane so that (5.7) converges to zero as $R \to \infty$. Thus, we have for $t < 0$ that

$$h_n(t) = \frac{1}{2\pi} \oint\limits_{\Gamma^+} \frac{-v_n \Re(\omega_n^+)^2}{(\omega - \omega_n^+)(\omega - \omega_n^-)} e^{-\mathrm{i}\omega t} \, d\omega = 0,$$

since the integrand is holomorphic in the upper half-plane. Conversely, if $t \geq 0$ then we should choose to integrate over Γ_R^- in order for (5.7) to disappear. Then, we see that for $t \geq 0$ it holds that

$$\begin{aligned} h_n(t) &= \frac{1}{2\pi} \oint\limits_{\Gamma^-} \frac{-v_n \Re(\omega_n^+)^2}{(\omega - \omega_n^+)(\omega - \omega_n^-)} e^{-\mathrm{i}\omega t} \, d\omega \\ &= \mathrm{i} \operatorname{Res}\left(\frac{-v_n \Re(\omega_n^+)^2}{(\omega - \omega_n^+)(\omega - \omega_n^-)} e^{-\mathrm{i}\omega t}, \omega_n^+ \right) + \mathrm{i} \operatorname{Res}\left(\frac{-v_n \Re(\omega_n^+)^2}{(\omega - \omega_n^+)(\omega - \omega_n^-)} e^{-\mathrm{i}\omega t}, \omega_n^- \right). \end{aligned}$$

We can simplify the expressions for the residues at the two simple poles to reach the result. □

Understanding the behavior of the system of coupled subwavelength resonators as a function of time allows us to examine other properties of the cochlea-like array. For example, the asymmetry of the spatial eigenmodes $u_n(x)$ means that the decomposition from Theorem 5.1.1 replicates the cochlea's famous traveling wave behavior, as was observed in Figure 3.7. That is, in response to an impulse the position of maximum amplitude moves slowly from the left to the right in the array. It should also be noted that the fact that $h_n(t) = 0$ for $t < 0$ ensures the causality of the modal expansion in Theorem 5.1.1.

With Theorem 5.1.1 in hand, we can see that our asymptotic results are not only useful for designing a physical structure that mimics the action of the cochlea in response to a sound wave, but also for describing approximately (in an asymptotic sense) how the system can be decomposed incoming sound waves. We propose using this as the basis for a biomimetic signal processing approach.

From Theorem 5.1.1, we know that the pressure field scattered by the cochlea-like array of resonators is described by a modal decomposition whose coefficients take the form of convolutions with the functions h_n. We wish to explore the properties of this decomposition, given for a sound $t \mapsto s(t)$ by

$$a_n[s](t) = (s * h_n)(t), \quad n = 1, \ldots, N, \tag{5.8}$$

which we will refer to as the *subwavelength scattering transform*. Convolutional signal processing algorithms such as this have been explored in detail [116, 119]. Here, we will present just a few elementary properties, to give some insight into the features of the algorithm that is deduced from our biomimetic approach.

Since the resonant frequencies all have negative imaginary parts, each h_n is a windowed oscillatory mode that acts as a band pass filter centered at $\Re(\omega_n^+)$. The frequency support of the 22 filters derived from our array of 22 resonators is shown in Figure 5.1.

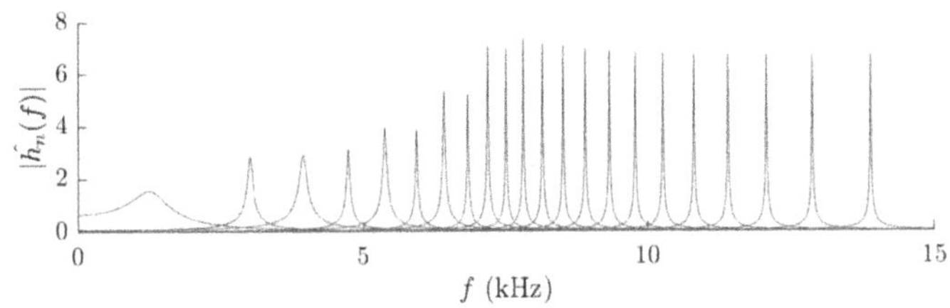

Figure 5.1: The frequency support of the band-pass filters h_n cover a range of frequencies that are audible to humans. Shown here for the case of 22 resonators. The resonators act as an array of band-pass filters, with properties (center frequency and bandwidth) determined by the corresponding resonant frequencies $\omega_n^{\pm} \in \mathbb{C}$.

Since the imaginary part of the lowest frequency is much larger than the others (see, e. g., Figure 4.1), h_1 acts somewhat as a low-pass filter.

The basis functions h_n take specific forms known as *gammatones*. A gammatone is a sinusoidal mode windowed by a gamma distribution,

$$g(t; m, \omega, \phi) = t^{m-1} e^{\Im(\omega)t} \cos(\Re(\omega)t - \phi), \quad t \geq 0, \tag{5.9}$$

for some order $m \in \mathbb{N}^+$ and constants $\omega \in \{z \in \mathbb{C} : \Im(z) < 0\}$, $\phi \in \mathbb{R}$. This is sketched in Figure 5.2. We notice that h_n is a first-order gammatone, i. e., $h_n(t) = c_n g(t; 1, \omega_n^+, \pi/2)$. Higher-order gammatones appear if the transform (5.8) is applied repeatedly in a cascade. The use of cascaded filters is a common approach both for designing auditory processing approaches [116] and convolutional networks in general [119]. The intuition is that a cascade allows the algorithm to resolve higher-order information contained in a signal. More details on this cascade and the emergence of higher-order gammatones can be found in the appendices of [6].

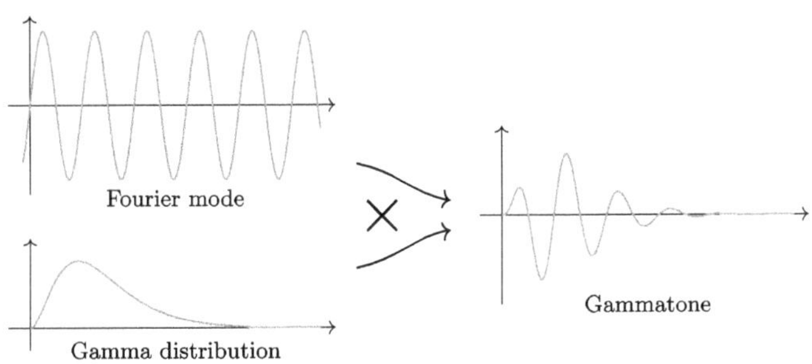

Figure 5.2: The cochlea-like metamaterial developed here has a response given by convolution with gammatones. A gammatone is an oscillating Fourier mode multiplied by a gamma distribution. This suggests using these gammatone functions as the basis for biomimetic signal processing algorithms.

The appearance of gammatones in this setting is promising as gammatones have been used widely in the literature to model auditory filters. This is because filters with gammatone kernels have been shown to approximate auditory function well, matching relatively well with physiological data and cochlear modeling [136, 87, 37, 115]. They are also relatively straightforward to analyze and implement [116], as we shall see below.

As well as agreeing with the trends in the bio-inspired signal processing literature, the gammatone decomposition we have derived has some advantageous continuity properties. The gammatone functions h_n are bounded and continuous, meaning that if $s \in L^1(\mathbb{R})$ then $s * h_n \in L^\infty(\mathbb{R})$. If, moreover, s is compactly supported then the decay properties of h_n mean that $s * h_n \in L^p(\mathbb{R})$ for any $p \in [1, \infty]$. Further, we have the following lemmas which characterize the continuity and stability of $s \mapsto s * h_n$.

Lemma 5.1.2 (Continuity of representation). *Consider the subwavelength scattering transform coefficients given by* (5.8). *There exists a positive constant C_1 such that for any $n \in \{1, \ldots, N\}$ and any signals $s_1, s_2 \in L^1(\mathbb{R})$ it holds that*

$$\|a_n[s_1] - a_n[s_2]\|_\infty \leq C_1 \|s_1 - s_2\|_1.$$

Proof. It holds that

$$C_1 := \sup_{n \in \{1,\ldots,N\}} \sup_{x \in \mathbb{R}} (1 - c) |h_n(x)| < \infty.$$

Then, the result follows from the fact that

$$|a_n[s_1](t) - a_n[s_2](t)| \leq \int_{-\infty}^{\infty} |s_1(u) - s_2(u)| |h_n(t - u)| \, du,$$

for any $t \in \mathbb{R}$. □

The continuity property stated in Lemma 5.1.2 implies, in particular, that the representation of a signal is stable with respect to additive noise. An additional useful property is for a representation to be stable with respect to time warping, i. e., with respect to composition with the operator $T_\tau f(t) = f(t + \tau(t))$, where τ is some appropriate function of time.

Lemma 5.1.3 (Pointwise stability to time warping). *Consider the subwavelength scattering transform coefficients given by* (5.8). *For $\tau \in C^0(\mathbb{R}; \mathbb{R})$, let T_τ be the associated time warping operator, given by $T_\tau f(t) = f(t + \tau(t))$. Then, there exists a positive constant C_2 such that for any $n \in \{1, \ldots, N\}$ and any signal $s \in L^1(\mathbb{R})$ it holds that*

$$\|a_n[s] - a_n[T_\tau s]\|_\infty \leq C_2 \|s\|_1 \|\tau\|_\infty.$$

Proof. Let h_n' denote the first derivative of h_n on $(0,\infty)$. Then, we see that

$$C_2 := \sup_{n\in\{1,\dots,N\}} \sup_{x\in(0,\infty)} |h_n'(x)| < \infty,$$

and, by the mean value theorem, that for $t \in \mathbb{R}$,

$$|h_n(t-\tau(t)) - h_n(t)| \le C_2|\tau(t)|.$$

Thus, we see that for any $t \in \mathbb{R}$,

$$\begin{aligned} |a_n[s] - a_n[T_\tau s]| &\le \int_{-\infty}^{\infty} |s(t-u)||h_n(u) - h_n(u-\tau(u))| \,\mathrm{d}u, \\ &\le C_2\|\tau(u)\|_\infty \int_{-\infty}^{\infty} |s(t-u)| \,\mathrm{d}u. \end{aligned}$$

□

We can improve on the notion of stability from Lemma 5.1.3 by taking temporal averages of the coefficients. A particular advantage of such an approach is that it gives outputs that are invariant to translation (cf. the motivation behind the design of the scattering transform [45, 118]). Let $\langle a_n[s]\rangle_{(t_1,t_2)}$ denote the average of $a_n[s](t)$ over the interval (t_1, t_2), given by

$$\langle a_n[s]\rangle_{(t_1,t_2)} = \frac{1}{t_2-t_1}\int_{t_1}^{t_2} a_n[s](t)\,\mathrm{d}t. \tag{5.10}$$

Lemma 5.1.4 shows that temporal averages are approximately invariant to translations if the length of the window is large relative to the size of the translation (i. e., if $t_2 - t_1 \gg \|\tau\|_\infty$).

Lemma 5.1.4 (Stability of averages to time warping). *Consider the subwavelength scattering transform coefficients given by* (5.8). *For $\tau \in C^1(\mathbb{R};\mathbb{R})$, let T_τ be the associated time warping operator, given by $T_\tau f(t) = f(t+\tau(t))$. Suppose that τ is such that $\|\tau'\|_\infty < \frac{1}{2}$. Then, there exists a positive constant C_3 such that for any $n \in \{1,\dots,N\}$, any time interval $(t_1,t_2) \subset \mathbb{R}$ and any signal $s \in L^1(\mathbb{R})$ it holds that*

$$|\langle a_n[s]\rangle_{(t_1,t_2)} - \langle a_n[T_\tau s]\rangle_{(t_1,t_2)}| \le C_3\|s\|_1\left(\frac{2}{t_2-t_1}\|\tau\|_\infty + \|\tau'\|_\infty\right).$$

Proof. Since $\|\tau'\|_\infty \le c < 1$, $\varphi(t) = t - \tau(t)$ is invertible and $\|\varphi'\|_\infty \ge 1 - c$, it holds that

$$\begin{aligned} \int_{t_1}^{t_2} (h_n(t-\tau(t)) - h_n(t))\,\mathrm{d}t &= \int_{\varphi(t_1)}^{\varphi(t_2)} h_n(t)\frac{1}{\varphi'(\varphi^{-1}(t))}\,\mathrm{d}t - \int_{t_1}^{t_2} h_n(t)\,\mathrm{d}t \\ &= \int_{I_1-I_2} h_n(t)\frac{1}{\varphi'(\varphi^{-1}(t))}\,\mathrm{d}t + \int_{t_1}^{t_2} h_n(t)\frac{\tau'(\varphi^{-1}(t))}{\varphi'(\varphi^{-1}(t))}\,\mathrm{d}t, \end{aligned}$$

for some intervals $I_1, I_2 \subset \mathbb{R}$, each of which has length bounded by $\|\tau\|_\infty$. Now, define the constant

$$C_3 := \sup_{n\in\{1,\dots,N\}} \sup_{x\in(0,\infty)} (1-c)^{-1}|h_n(x)| < \infty.$$

Finally, we can compute

$$\begin{aligned}
&\langle a_n[s]\rangle_{(t_1,t_2)} - \langle a_n[T_\tau s]\rangle_{(t_1,t_2)} \\
&= \frac{1}{t_2-t_1}\int_{-\infty}^{\infty} s(u) \int_{t_1}^{t_2} (h_n(t-u-\tau(t)) - h_n(t-u))\,dt\,du \\
&= \frac{1}{t_2-t_1}\int_{-\infty}^{\infty} s(u)\Bigg(\int_{I_1-I_2} h_n(t-u)\frac{1}{\varphi'(\varphi^{-1}(t-u))}\,dt \\
&\quad + \int_{t_1}^{t_2} h_n(t-u)\frac{\tau'(\varphi^{-1}(t-u))}{\varphi'(\varphi^{-1}(t-u))}\,dt\Bigg)\,du,
\end{aligned}$$

meaning that

$$|\langle a_n[s]\rangle_{(t_1,t_2)} - \langle a_n[T_\tau s]\rangle_{(t_1,t_2)}| \le \frac{1}{t_2-t_1}\|s\|_1[2\|\tau\|_\infty C_3 + (t_2-t_1)C_3\|\tau'\|_\infty]. \qquad \square$$

We have seen that the structure proposed by Theorem 5.1.1, of an array of convolutions with gammatones, provides an excellent starting point for a signal processing architecture. This result, which was derived from the cochlea-like metamaterial, matches direct observations of the human auditory system and gives robust, stable representations of signals. In the rest of this chapter, we will use some additional observations of biological sensing systems to propose additional processing steps that could be used to augment this simple algorithm.

5.2 Natural sounds

The human auditory system does more than just extract information from a signal locally in time, it is also able to recognize global properties of a sound and can appreciate notions of timbre and quality. We would like to design an approach that can account for this, by adding an additional processing step to the subwavelength scattering transform that was derived above. The approach set out in this section is tailored to the class of *natural sounds*. This is, a class of natural and behaviorally-significant sounds to which humans are known to be adapted. These sounds have observed statistical properties that we are able to exploit.

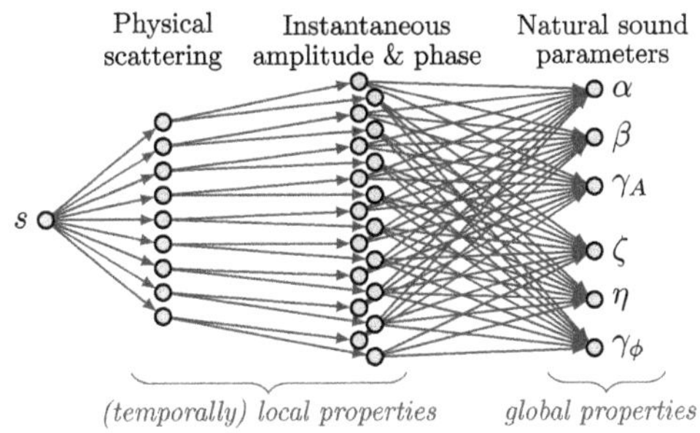

Figure 5.3: The architecture proposed to encode sounds using their "natural sound coefficients" first applies filters derived from physical scattering by a cochlea-like device, then extracts the instantaneous amplitude and phase before, finally, estimating the parameters of the associated natural sound distributions.

Let us briefly summarize what has been observed about the low-order statistics of natural sounds [30, 31, 157, 162]. For a sound $s(t)$, let $a_\omega(t)$ be the component at frequency ω (obtained, e. g., through the application of a band-pass filter centered at ω). Then we can write that

$$a_\omega(t) = A_\omega(t)\cos(\omega t + \phi_\omega(t)), \tag{5.11}$$

where $A_\omega(t) \geq 0$ and $\phi_\omega(t)$ are the instantaneous amplitude and phase, respectively. We view $A_\omega(t)$ and $\phi_\omega(t)$ as stochastic processes and wish to understand their statistics.

The most famous characteristic of natural sounds is that several properties of their frequency components vary according to the inverse of the frequency. In particular, it is well known that the power spectrum (the square of the Fourier transform) of the amplitude satisfies a relationship of the form

$$S_{A_\omega}(f) = |\hat{A}_\omega(f)|^2 \propto \frac{1}{f^\gamma}, \quad 0 < f < f_{\max}, \tag{5.12}$$

for a positive parameter γ (which often lies in a neighborhood of 1) and some maximum frequency $f_{\max}$. Further, this property is independent of ω, i. e., of the frequency band that is studied [30].

Consider the log-amplitude, $\log_{10} A_\omega(t)$. It has been observed that for a variety of natural sounds (including speech, animal vocalizations, music, and environmental sounds) the log-amplitude is locally stationary, in the sense that it satisfies a statistical distribution that does not depend on the time t. Suppose we normalize the log-amplitude so that it has zero mean and unit variance, giving a quantity that is invariant to amplitude scaling. Then, the normalized log-amplitude averaged over some time interval $[t_1, t_2]$ satisfies a distribution of the form [31]

$$p_A(x) = \beta \exp(\beta x - \alpha - e^{\beta x - \alpha}), \tag{5.13}$$

where α and β are real-valued parameters and $\beta > 0$. Further, this property does not depend on the frequency band and is scale invariant in the sense that it is independent of the time interval over which the average is taken.

The properties of the instantaneous phase $\phi_\omega(t)$ have also been studied. Similar to the instantaneous amplitude, the power spectrum S_{ϕ_ω} of the instantaneous phase satisfies a $1/f$-type relationship, of the same form as (5.12). On the other hand, the instantaneous phase is non-stationary (even locally), making it difficult to describe through the above methods. A more tractable quantity is the instantaneous frequency, defined as

$$\lambda_\omega = \frac{\mathrm{d}\phi_\omega}{\mathrm{d}t}.$$

It has been observed that $\lambda_\omega(t)$ is locally stationary for natural sounds and the temporal mean of its modulus satisfies a distribution p_λ of the form [30]

$$p_\lambda(x) \propto \left(\zeta^2 + x^2\right)^{-\eta/2}, \tag{5.14}$$

for positive parameters ζ and $\eta > 1$.

For a given natural sound, we wish to find the parameters that characterize its global properties, according to (5.12)–(5.14). Given a signal s, we first compute the convolution with the band-pass filter h_n to yield the spectral component at the frequency $\Re(\omega_n^+)$, given by

$$a_n[s](t) = A_n(t)\cos(\Re(\omega_n^+)t + \phi_n(t)).$$

We extract the functions A_n and ϕ_n from $a_n[s]$ using the Hilbert transform [30, 78, 42]. In particular, we have that

$$a_n[s](t) + \mathrm{i}H(a_n[s])(t) = a_n[s](t) + \frac{\mathrm{i}}{\pi}\int_{-\infty}^{\infty} \frac{a_n[s](u)}{t-u}\,du = A_n(t)e^{\mathrm{i}(\Re(\omega_n^+)t + \phi_n(t))},$$

from which we can extract A_n and ϕ_n by taking the complex modulus and argument, respectively. It is not obvious that the Hilbert transform $H(a_n[s])$ is well defined. Indeed, we must formally take the principal value of the integral. For a signal that is integrable and has compact support, $H(a_n[s])(t)$ exists for almost all $t \in \mathbb{R}$.

Given the functions A_n and ϕ_n, the power spectra $S_{A_n}(f)$ and $S_{\phi_n}(f)$ can be computed by applying the Fourier transform and squaring. We estimate the relationships of the form (5.12) by first averaging the N power spectra, to give $\overline{S}_A(f) := \frac{1}{N}\sum_n S_{A_n}(f)$ and $\overline{S}_\phi(f) := \frac{1}{N}\sum_n S_{\phi_n}(f)$ before fitting curves $f^{-\gamma_A}$ and $f^{-\gamma_\phi}$ using least-squares regression.

We estimate the parameters of the probability distributions (5.13) and (5.14) by normalizing both $\log_{10} A_n(t)$ and $\lambda_n(t)$ so that

$$\langle \log_{10} A_n \rangle = 0, \quad \langle (\log_{10} A_n)^2 \rangle = 1,$$

and similarly for $\lambda_n(t)$, before repeatedly averaging the normalized functions over intervals $[t_1, t_2] \subset \mathbb{R}$. Curves of the form (5.13) and (5.14) are then fitted to the resulting histograms (which combine the temporal averages from different filters $n = 1, \dots, N$ and different time intervals $[t_1, t_2]$) using nonlinear least-squares optimization. A schematic of this parameter extraction architecture is given in Figure 5.3. An example of the four datasets and their fitted distributions are shown in Figure 5.4 for a short recording of a trumpet playing a single note. Table 5.1 shows some other examples of these parameters for various natural sounds.

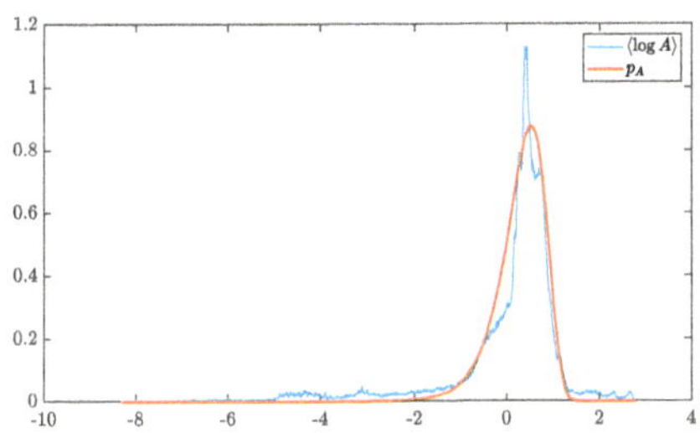

(a) A histogram of the normalised time-averaged log-amplitude, fitted to a distribution of the form (5.13).

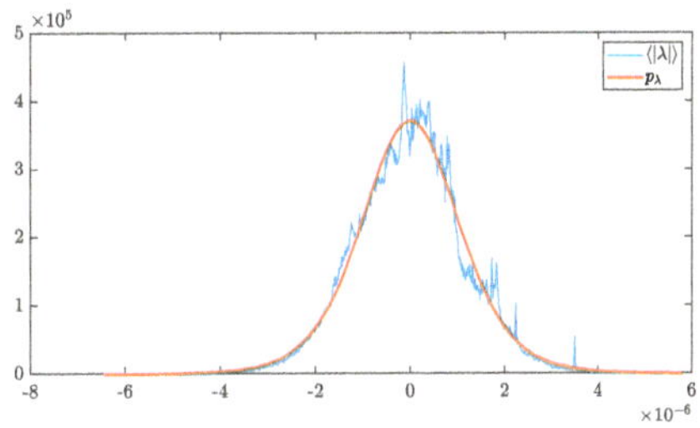

(b) A histogram of the normalised time-averaged instantaneous frequency, fitted to a distribution of the form (5.14).

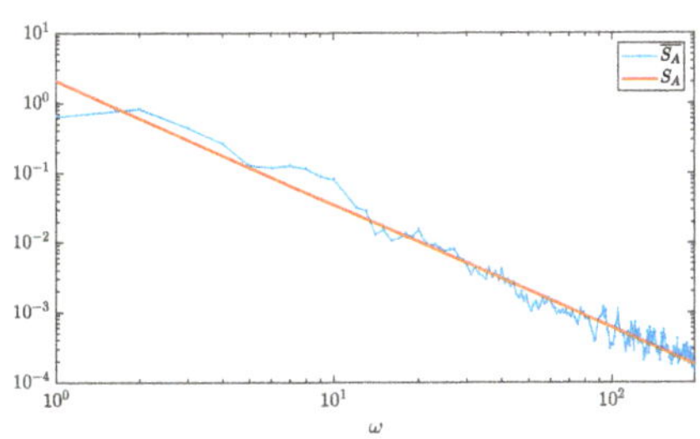

(c) The averaged (over different frequency bands) power spectra of the instantaneous amplitude, fitted to a distribution of the form (5.12).

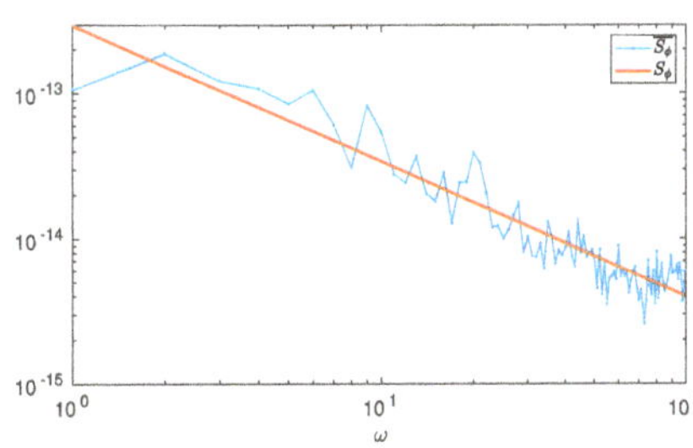

(d) The averaged (over different frequency bands) power spectra of the instantaneous phase, fitted to a distribution of the form (5.12).

Figure 5.4: Natural sounds have been observed to satisfy given low-order statistical properties. The output of scattering by a cochlea-like metamaterial will satisfy known distributions, as a result. Here, the data extracted by the cochlea-like device from a recording of a trumpet playing a single note are shown along with the fitted distributions.

Table 5.1: Values of the estimated distribution parameters for different samples of natural sounds; γ_A and γ_ϕ capture the $f^{-\gamma}$ relationships of the averaged power spectra $\bar{S}_A$ and $\bar{S}_\phi$. The distribution p_A of the time-averaged, normalized log-amplitude is parametrized by α and β, while ζ and η parametrize the distribution p_λ of the time-averaged instantaneous frequency.

	trumpet	violin	cello	thunder	baby speech	adult speech	running water	crow call
γ_A	1.767	1.563	1.528	1.415	1.763	1.808	1.466	1.571
α	1.244	0.375	0.284	0.474	0.517	0.528	0.336	0.649
β	2.390	0.783	0.841	0.596	0.747	0.894	0.484	0.896
γ_ϕ	0.763	0.871	0.6977	0.446	1.192	1.125	1.088	0.908
ζ ($\times 10^{-6}$)	2.878	3.433	6.1149	6.322	4.773	5.176	5.200	4.212
η	8.579	11.824	8.679	8.315	9.660	9.358	9.290	10.475

The observed properties of natural sounds give us six real-valued coefficients γ_A, α, β, γ_ϕ, ζ, and η that portray global properties of a sound. Given the output from the cochlea-like metamaterial, we are able to extract these parameters (see Figure 5.3). Global parameters of this kind have been shown to capture, in a perceptually significant sense, the quality of a sound and play an important role in our ability to recognize sounds efficiently [123, 168, 111, 132]. Adding these parameters to the information already gained from the subwavelength scattering transform can improve the extent to which the representation algorithm is able to mimic the perceptual abilities of the human auditory system.

As a very simple demonstration of the value of these natural sound coefficients, we perform a simple classification experiment. We took recordings of ten different musical instruments and used the natural sound coefficients as a low-dimensional representation for classification. First, a dictionary of reference values was created. For this, we used ten samples of each instrument and computed the average values of each of the six coefficients. These values are shown in Table 5.2. These values were then used as the reference values for classifying unknown recordings. An additional collection of musical recordings was used as a test set. For each recording in the test set, the six natural sound coefficients were computed and compared to the dictionary. The L^1 norm on $\mathbb{R}^6$ was used to compare each sample with the dictionary, with the categorization being assigned according to which entry in the dictionary is closest to the data point.

The results from the musical instrument classification test are shown in Table 5.3. Each row contains the data for recordings of a given instrument, with the values in each column indicating how often the recordings of that instrument were classified in each category. We can see that for some instruments, such as the flute, the tuba and the cello, most of the recordings are correctly classified. This clearly implies that the natural sound coefficients contain some meaningful information about sound and its inherent properties. Conversely, for some other instruments, such that the clarinet and the trombone, the success rate is much lower. It is not clear if this is an inherent property

Table 5.2: A "dictionary" of values of the estimated distribution parameters for different samples of musical instruments; γ_A and γ_ϕ capture the $f^{-\gamma}$ relationships of the averaged power spectra $\bar{S}_A$ and $\bar{S}_\phi$. The distribution p_A of the time-averaged, normalized log-amplitude is parametrized by α and β, while ζ and η parametrize the distribution p_λ of the time-averaged instantaneous frequency.

	bassoon	clarinet	flute	tuba	trombone	trumpet	cello	violin	piano	vibraphone
γ_A	1.65	1.86	2.01	1.59	1.81	1.84	1.80	1.81	1.62	1.93
α	1.90	3.33	3.59	1.34	1.53	1.84	8.17	1.79	1.87	2.23
β	3.62	5.41	6.71	3.32	3.87	5.60	17.2	3.60	0.67	5.39
γ_ϕ	0.55	1.16	1.52	1.13	0.96	1.33	0.97	1.29	0.72	0.96
ζ ($\times 10^{-6}$)	1.59	4.73	2.21	2.98	1.88	1.71	4.04	2.05	2.46	3.42
η	22.2	15.5	13.1	9.16	12.6	18.6	5.69	16.3	43.7	69.0

Table 5.3: Results of using the dictionary from Table 5.2 to classify unknown recordings of music. Each row corresponds to the true instrument in the recording, with the values in each column indicating the number of times recordings of that instrument were classified in the given instrument category. The diagonal entries indicate successful classification. The overall success rate in this test was 31.1%.

		Classified instrument									
		bassoon	clarinet	flute	tuba	trombone	trumpet	cello	violin	piano	vibraphone
True instrument	bassoon	**2**	0	3	5	4	1	0	2	0	2
	clarinet	0	**4**	17	3	0	2	21	3	0	2
	flute	1	2	**24**	0	0	6	7	3	0	4
	tuba	1	0	4	**16**	1	2	2	0	1	0
	trombone	3	0	4	5	**1**	2	1	5	0	5
	trumpet	1	0	8	3	1	**7**	2	0	1	4
	cello	1	2	1	5	0	1	**22**	0	0	1
	violin	0	8	6	13	6	2	7	**16**	6	4
	piano	3	0	0	13	3	0	7	0	**23**	4
	vibraphone	3	2	1	3	2	2	5	3	12	**8**

of these instruments or if it is due to poorly calibrated training data; the dictionary was generated using a relatively small collection of ten samples, so is liable to be heavily influenced by errors or outliers. The overall success rate for this simple experiment was 31.1 % which, despite being well below the performance of modern signal processing approaches, is enough to indicate that these natural sound coefficients contain some meaningful information about the sound.

In a sense, it is remarkable that this simple algorithm can recover any meaningful success rate in a classification problem. Firstly, it is reasonable to expect that a vector with just six entries cannot hope to describe the rich variety of musical sounds that exist. Secondly, the natural sounds coefficients do not contain any temporal information, as they are global statistical quantities. Some sounds are largely recognized by how they evolve over time. A good example of this in our setting is a piano, where the sound is created by a hammer striking a tightened string. As a result, the sound of a piano is characterized by having a distinctive initial strike followed by steady decay.[1] An interpretation of is that the natural sound coefficients are partly describing what a musician would call the *timbre* of the sound: its character or quality. Since these coefficients require minimal computational expense to compute, we suggest that they could be added to existing classification routines, with the hope of improving classification accuracy and robustness.

5.3 Random projections

It is also possible to look beyond the realm of human hearing for inspiration on effective signal processing approaches. For example, it is often the case that connections between neurons are formed at random and they differ between different organisms. The fruit fly's olfactory system is an example that is particularly well understood [60, 149, 150] and there is also some evidence that the human brain behaves similarly [149]. An approach of this type has the advantage that it is not necessary to undertake expensive training periods in order to learn the connections. Instead, the algorithm randomly forms many connections and selects the most useful ones.

We will briefly summarize the key properties of the fruit fly's olfactory system below. For more details, we refer the interested reader to [44, 60, 149, 150]. This will inspire us to form a two-step algorithm for processing a feature vector. This feature vector could be obtained by a subwavelength scattering transform of the form (5.8), for example. The first step will be to project the feature vector into a high-dimensional space by multiplying by a random matrix. This matrix will be rectangular, such that the dimension of the image space is much larger than that of the initial feature vector. Following this, we will apply a "cap" operation, that sets all but the few largest entries in the vector to zero. This means the final representation can be chosen to be suitably sparse.

Random projections have been used in a variety of computational applications, often with the aim of reducing computational time. This is particularly the case when they are used to replace expensive training steps. For example, they have been used in neural networks to randomly select weights [28] or features [140]. In both cases, there is little loss of performance with a significant reduction in computation time. Similarly, random projections have been used to reduce dimensionality, either as the first step in a classification algorithm [27] or to approximate kernel functions [139]. In many cases, the structure of the random projections is specifically chosen for the task in hand (e. g., to detect corners in images, as in [27]). Conversely, the method developed in this work

1 For example, piano music played backwards does not bear much familiarity with piano music as we know it.

will be, theoretically, independent of the setting and will treat the problem of classifying a given feature vector of arbitrary dimension.

As well as reducing computation time, there are also deep connections between the use of random projections and classification robustness. For example, [28] shows that more robust targets (in a suitable sense) can be more effectively compressed by being randomly projected to a lower dimensional space. As a result, concepts that are sufficiently robust can be successfully randomly projected to a lower dimension, where they can be classified. In the example we will present below, we will explore the extent to which our bio-inspired algorithm facilitates robust classification. In particular, we will show that, for the problem of musical genre classification, the use of random projections gives an improvement in classification accuracy when random noise is added to the signal.

Early processing of odors in the fruit fly's olfactory system consists of roughly three steps. In the first step, olfactory receptor neurons (ORNs) located in the fly's antennae detect an odor and send a signal to projector neurons in the antennal lobe. In the second step, the projector neurons transmit the firing rates to parts of the fly's brain known as the mushroom body and the lateral horn. We will focus on the mushroom body, since this part of the fly's brain is known to be important for learning new smells and creating memories associated with them [121]. Here, signals are transmitted randomly to a large number of Kenyon cells. Finally, in the third step, anterior paired lateral neurons suppress a large number of the Kenyon cells, so that only those with the highest firing rates are uninhibited. A sketch of the important connections in the fly's olfactory system is shown in Figure 5.5.

After being activated, ORNs fire to structures called glomeruli in the antennal lobe. Each glomerulus receives the input from all ORNs of a particular type; there are there-

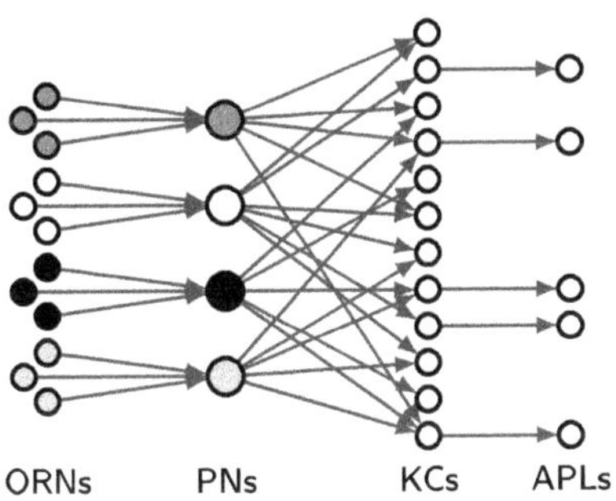

Figure 5.5: The first steps of the fruit fly's olfactory system, shown as a simplified schematic. Odors are detected by olfactory receptor neurons (ORNs) in the antennae and the maxillary palps. ORNs of the same type fire to the same projector neurons (PNs). Projector neurons then fire to the mushroom body, where signals are transmitted in random combinations to a large number of Kenyon cells (KCs). Finally, anterior paired lateral (APL) neurons inhibit the output of 95% of the KCs, leaving only those with the largest firing rates.

fore about 50 glomeruli. In the glomerulus, ORNs make synaptic contact with a projector neuron. At this stage, the odor information thus can be represented as a 50-dimensional vector, where each coefficient corresponds to the firing rate of a single type of ORN. That signal is then projected to 2000 Kenyon cells in the mushroom body, resulting in a 40-fold increase in signal dimension. Each Kenyon cell receives the firing rates of approximately 6 projector neurons and sums them up [60]. Crucially, the projection to Kenyon cells is random, in the sense that the latter do not receive a signal from fixed projector neurons depending on the type of smell detected. From one fly to another, a similar smell triggers different Kenyon cells, even if the same types of ORNs have been activated.

Evidence suggests that the set of Kenyon cells activated after exposure to an odor forms the odor "tag" that allows the fly to recognize it: [47] shows that a large overlap in the firing rates of a group of Kenyon cells is a good predictor of whether a fly will judge two smells to be similar. Moreover, the results of [47] show that the entire population of Kenyon cells is not necessary to discriminate smells, but rather that a subset of 25 cells gives sufficient information to predict the fly's response. This is a consequence of the fact that glomeruli fire randomly to Kenyon cells and the results are summed. Thus, the entire information they project can be found in a relatively small subset of the Kenyon cells. One can then wonder why the information is spread over 2000 Kenyon cells, when many fewer seem to provide enough information. Stevens [149] argues that the reason for such a large, redundant representation of smells in the mushroom body is to provide multiple representations, so that the fly can later utilize the one containing the crucial information.

Finally, the last step in the fly's olfactory system consists of an inhibitory process. Anterior paired lateral (APL) neurons deactivate about 95 % of Kenyon cells, leaving only those with the highest firing rates [60, 149]. As a result, the final vector representation of the smell information is relatively large (a vector with 2000 entries) and very sparse. This is easy to reproduce with a cap operation, that acts in the same way by setting all but the largest entries to zero.

Altogether, those steps amount to a random projection of the initial 50-dimensional vector into a 2000-dimensional space, followed by a nonlinear operation that only keeps a fixed number of the highest coefficients. The initial 50-dimensional signal vector contains the firing rates of each type of ORN. The projection of the glomeruli to the Kenyon cells can be described by multiplication by a random matrix with size 2000×50 and entries drawn from $\{0, 1\}$. Each row corresponds to a particular Kenyon cell; for each glomerulus that fires a signal to that cell, we write a 1 in the corresponding column of the row vector. All other entries are set to 0. As only few of the projector neurons fire to the next step, the random matrix should be sparse. We modify this formulation slightly in our algorithm, to take advantage of other beneficial properties (e. g., when the entries have a symmetric distribution).

The application of a random projection followed by an inhibitory process is not unique to the fly's olfactory system. In fact, similar processes play a role in three parts of the brain: the cerebellum, the hippocampus, and the olfactory system. Stevens [149] explains that the way those three structures process information follows a similar three-step architecture to the fly's olfactory system. In the first stage, the information arriving from other brain areas is assembled into a neural code. In the second stage, that code is passed on to a greatly enlarged number of neurons. Finally, in the third stage, this code is broken down to be interpreted in further information processing steps. The corresponding models for the human brain are somewhat more complicated than the model for the fruit fly, as depicted in Figure 5.5. For this reason, we will restrict ourselves to the fruit fly case here, but we refer the interested reader to [149, 134].

Motivated by the above discussion, we will consider the transformation $A : \mathbb{R}^m \to \mathbb{R}^n$ given by

$$A(s) = c_k(Ms), \tag{5.15}$$

where $M \in \mathbb{R}^{n\times m}$ is a random matrix and $c_k : \mathbb{R}^n \to \mathbb{R}^n$ is a cap operation that keeps the k largest (in magnitude) entries of a vector and sets all the others to zero. Implicitly, we need $k \leq n$. On top of this, we will focus on the case $n \gg m$, to replicate the random projection from few projector neurons to many Kenyon cells in the fly's olfactory system. We will choose $M \in \mathbb{R}^{n\times m}$ to be such that it is a random matrix whose entries m_{ij} are independent and identically distributed and given by the difference between two independent Bernoulli random variables:

$$m_{ij} = X_{ij} - Y_{ij}, \quad i = 1, \dots, n, j = 1, \dots, m, \tag{5.16}$$

where X_{ij} and Y_{ij} are independent Bernoulli random variables with parameter $p \in (0, 1)$:

$$X_{ij}, Y_{ij} = \begin{cases} 1 & \text{with probability } p, \\ 0 & \text{with probability } 1 - p. \end{cases} \tag{5.17}$$

This choice of random matrix is motivated by the way that, in the fly's olfactory system, each Kenyon cell receives the firing rates of multiple projector neurons and sums them up. However, we have added the extra feature that each entry of M should have mean zero, which will yield several useful mathematical properties using the existing literature on properties of random matrices. Similar matrices with symmetric distributions were considered. e. g., by [2, 27, 28, 82]. We will choose $p \in (0, 1)$ to be small, such that M is likely to be sparse. This will improve the speed of calculations, especially when M is a very large matrix.

Intuitively, we would like our transformation A to satisfy certain characteristics. Firstly, we want it to be continuous in the sense that similar signals should still be close after transformation. On the other hand, we want signals that are different enough to be further apart after being transformed by A. While the latter is a bit more complex to guarantee, we will be able to show that, with high probability, our transformation preserves similarities between vectors.

We begin by exploring the properties of multiplication by the random matrix M. The following lemma describes the distribution of the entries of the matrix M.

Lemma 5.3.1. *If $m_{ij} \sim X - Y$, where X, Y are Bernoulli random variables with parameter $p \in (0,1)$, then it holds that $\mathbb{E}(m_{ij}) = 0$, $\mathbb{P}(m_{ij} = 0) = 2p^2 - 2p + 1$ and $\mathrm{Var}(m_{ij}) = 2p(1-p)$.*

Proof. Let $Z = X - Y$ where X, Y are Bernoulli random variables with parameter $p \in (0,1)$. Then, the expectation follows by a simple calculation: $\mathbb{E}(Z) = \mathbb{E}(X - Y) = \mathbb{E}(X) - \mathbb{E}(Y) = p - p = 0$. Similarly, we can calculate that $\mathbb{P}(Z = 0) = \mathbb{P}(X = 0, Y = 0) + \mathbb{P}(X = 1, Y = 1) = (1-p)^2 + p^2$. For the variance, we note that since $Z \in \{-1, 0, 1\}$, we have $Z^2 \in \{0, 1\}$. Thus $\mathbb{E}(Z^2) = \mathbb{P}(Z^2 = 1) = 1 - \mathbb{P}(Z = 0) = 2p(1-p)$. Since $\mathbb{E}(Z) = 0$, we have $\mathrm{Var}(Z) = \mathbb{E}(Z^2) = 2p(1-p)$. □

We first present a theorem that bounds the operator norm of the random matrix M with high probability.

Theorem 5.3.1. *Given the matrix $M \in \mathbb{R}^{n\times m}$, whose entries are each the difference of independent and identically distributed Bernoulli random variables, there exist real-valued constants C and $c > 0$ such that*

$$\mathbf{P}(\|M\|_{\mathrm{op}} > D\sqrt{n}) \leq C\exp(-cDn),$$

for all $D \geq C$. In particular, we have $\|M\|_{\mathrm{op}} = O(\sqrt{n})$ with probability that is exponentially close to 1.

Proof. This theorem and its proof can be found in [155, Theorem 2.1.3], for the more general case where M is any matrix with entries that are independent and identically distributed, have zero mean and are uniformly bounded by 1. From Lemma 5.3.1, we can see that this holds for our specific choice of M. □

This theorem gives us a bound on

$$\|Mx\| \leq \|M\|_{\mathrm{op}}\|x\|, \tag{5.18}$$

which guarantees that, with high probability, a vector's norm will not blow up due to multiplication by the random matrix M, even as the dimension of M becomes very large. Throughout this chapter, we will use $\|\cdot\|$ to denote the Euclidean norm (i. e., $\|\cdot\|_2$). In particular, it holds that

$$\|Mx - My\| \leq \|M\|_{\mathrm{op}}\|x - y\| \tag{5.19}$$

for any $x, y \in \mathbb{R}^m$.

We will however be able to give further bounds on $\|Mx - My\|$ by modifying a famous result by Johnson and Lindenstrauss, which states that a set of points in $\mathbb{R}^d$ can be mapped to $\mathbb{R}^k$ while approximately preserving distances between pairs of points, as long as k is large enough. Most of the literature focuses on the case where $k < d$, as this allows for data compression; however, this lemma is still informative to us, as it shows that we can map data points to a different dimension while more or less retaining their pairwise distances. Note that the standard formulation of this lemma only states that such a mapping exists, and does not specify what it might look like. The Johnson–Lindenstrauss lemma is a standard result that can be found in [2, Lemma 1.1], for example.

Theorem 5.3.2 (Johnson–Lindenstrauss lemma). *Given $\epsilon > 0$ and an integer n, let k be a positive integer such that $k \geq k_0 = O(\epsilon^{-2}\log n)$. For every set P of n points in $\mathbb{R}^d$, there exists $f : \mathbb{R}^d \to \mathbb{R}^k$ such that for all $u, v \in P$,*

$$(1-\epsilon)\|u-v\|^2 \leq \|f(u)-f(v)\|^2 \leq (1+\epsilon)\|u-v\|^2.$$

The next theorem shows that multiplication by a random matrix R, whose entries follow a distribution that is symmetric around zero and has unit variance, preserves the norm of vectors up to a scaling constant, with high probability. This constant depends on $\sqrt{n}$, where n is the dimension of the space into which $M \in \mathbb{R}^{n\times m}$ projects. The theorem and its proof come from [28, Theorem 1].

Theorem 5.3.3. *Let $R \in \mathbb{R}^{n\times m}$ be a random matrix, with each entry r_{ij} chosen independently from a distribution that is symmetric about the origin an has $\mathbb{E}(r_{ij}^2) = 1$.*
(i) *Suppose $B = \mathbb{E}(r_{ij}^4) < \infty$. Then, for any $\epsilon > 0$,*

$$\mathbb{P}\left(\left\|\frac{1}{\sqrt{n}}Ru\right\|^2 \leq (1-\epsilon)\|u\|^2\right) \leq \exp\left(-\frac{(\epsilon^2-\epsilon^3)n}{2(B+1)}\right) \quad \textit{for all } u \in \mathbb{R}^m;$$

(ii) *Suppose $\exists L > 0$ such that for any integer $k > 0$, $\mathbb{E}(r_{ij}^{2k}) \leq \frac{(2k)!}{2^k k!}L^{2k}$. Then, for any $\epsilon > 0$,*

$$\mathbb{P}\left(\left\|\frac{1}{\sqrt{n}}Ru\right\|^2 \geq (1+\epsilon)L^2\|u\|^2\right) \leq \exp\left(-(\epsilon^2-\epsilon^3)\frac{n}{4}\right) \quad \textit{for all } u \in \mathbb{R}^m.$$

As shown in Lemma 5.3.1, our matrix M satisfies $\mathbb{E}(m_{ij}^2) = 2p(1-p) \neq 1$. Therefore, we need the following corollary to extend Theorem 5.3.3 to settings with matrix entries drawn from a distribution with rescaled variance.

Corollary 5.3.1. *Let $M \in \mathbb{R}^{n\times m}$ be a random matrix whose entries m_{ij} are sampled independently and randomly from a distribution that is symmetric around the origin with $\mathbb{E}(m_{ij}^2) = \sigma^2 > 0$.*
(i) *Suppose $B = \mathbb{E}(m_{ij}^4) < \infty$. Then, for any $\epsilon > 0$,*

$$\mathbb{P}\left(\left\|\frac{1}{\sqrt{n}}Mu\right\|^2 \leq \sigma^2(1-\epsilon)\|u\|^2\right) \leq \exp\left(-\frac{(\epsilon^2-\epsilon^3)n}{2(\frac{1}{\sigma^4}B+1)}\right) \quad \textit{for all } u \in \mathbb{R}^m;$$

(ii) *Suppose* $\exists L > 0$ *such that for any integer* $k > 0$, $\mathbb{E}(m_{ij}^{2k}) \le \sigma^{2k} \frac{(2k)!}{2^k k!} L^{2k}$. *Then, for any* $\epsilon > 0$,

$$\mathbb{P}\left(\left\|\frac{1}{\sqrt{n}} Mu\right\|^2 \ge \sigma^2(1+\epsilon)L^2\|u\|^2\right) \le \exp\left(-(\epsilon^2 - \epsilon^3)\frac{n}{4}\right) \quad \textit{for all } u \in \mathbb{R}^m.$$

Proof. (i) Let R be the $n \times m$ matrix defined as $R := \frac{1}{\sigma} M$. Then clearly the entries $r_{ij} = \frac{1}{m_{ij}}$ of R are sampled from a distribution symmetric around 0. Moreover, $\mathbb{E}(r_{ij}^2) = \mathbb{E}(\frac{1}{\sigma^2} m_{ij}^2) = \frac{\sigma^2}{\sigma^2} = 1$. Finally, $\mathbb{E}(r_{ij}^4) = \mathbb{E}(\frac{1}{\sigma^4} m_{ij}^4) = \frac{1}{\sigma^4} B < \infty$. Hence R satisfies the conditions of Theorem 5.3.3, and we have

$$\mathbb{P}\left(\left\|\frac{1}{\sqrt{n}} Ru\right\|^2 \le (1-\epsilon)\|u\|^2\right) \le \exp\left(-\frac{(\epsilon^2 - \epsilon^3)n}{2(\frac{1}{\sigma^4} B + 1)}\right). \tag{5.20}$$

The left hand side can be rewritten as

$$\mathbb{P}\left(\left\|\frac{1}{\sqrt{n}}\frac{1}{\sigma} Mu\right\|^2 \le (1-\epsilon)\|u\|^2\right) = \mathbb{P}\left(\left\|\frac{1}{\sqrt{n}} Mu\right\|^2 \le \sigma^2(1-\epsilon)\|u\|^2\right). \tag{5.21}$$

(ii) Once again, let $R = \frac{1}{\sigma} M$. Clearly, for every integer k, we have

$$\mathbb{E}(r_{ij}^{2k}) = \mathbb{E}\left(\frac{1}{\sigma^{2k}} m_{ij}^{2k}\right) \le \frac{2k!}{2^k k!} L^{2k}. \tag{5.22}$$

Hence R satisfies the conditions of Theorem 5.3.3, and we have

$$\mathbb{P}\left(\left\|\frac{1}{\sqrt{n}} Ru\right\|^2 \ge (1+\epsilon)L^2\|u\|^2\right) \le \exp\left(-(\epsilon^2 - \epsilon^3)\frac{n}{4}\right). \tag{5.23}$$

The left hand side of this equation is equal to

$$\mathbb{P}\left(\left\|\frac{1}{\sqrt{n}}\frac{1}{\sigma} Mu\right\|^2 \ge (1+\epsilon)L^2\|u\|^2\right) = \mathbb{P}\left(\left\|\frac{1}{\sqrt{n}} Mu\right\|^2 \ge \sigma^2(1+\epsilon)L^2\|u\|^2\right),$$

giving the desired inequality. □

Finally, we can use these results to prove an analogous theorem about the effect of multiplying by the random matrix M.

Theorem 5.3.4. *Given the matrix* $M \in \mathbb{R}^{n \times m}$, *whose entries are each the difference of independent and identically distributed Bernoulli random variables with parameter* $p \in (0,1)$, *it holds for any* $\epsilon > 0$ *that*

$$\mathbb{P}\left((1-\epsilon)\|u - v\|^2 \le \frac{1}{n\sigma^2}\|Mu - Mv\|^2 \le (1+\epsilon)\|u - v\|^2\right) \ge 1 - e^{-(\epsilon^2 - \epsilon^3)\frac{n}{4}} - e^{-\frac{(\epsilon^2 - \epsilon^3)n}{2(\frac{1}{\sigma^2}+1)}},$$

for all $u, v \in \mathbb{R}^m$, *where* $\sigma^2 = 2p(1-p)$.

Proof. First, we need to show that our matrix M satisfies the conditions of Corollary 5.3.1. It was shown above that $\mathbb{E}(m_{ij}^2) = 2p(1-p)$. When $p \in (0,1)$, it holds that $2p(1-p) > 0$. Since $m_{ij}^2 \in \{0,1\}$, we have that $(m_{ij}^2)^k = m_{ij}^2$ for any integer k. Therefore, $\mathbb{E}(m_{ij}^{2k}) = \mathbb{E}(m_{ij}^2) = 2p(1-p)$ for all positive integers k. Picking $L = [2p(1-p)]^{-1/2}$ gives the relation

$$\mathbb{E}(m_{ij}^{2k}) = 2p(1-p) \le 1 \le [2p(1-p)]^k \frac{(2k)!}{2^k k!} L^{2k}, \tag{5.24}$$

so that the condition in the second part of Corollary 5.3.1 is satisfied.

Finally, we can apply Corollary 5.3.1 to the vector $(u-v)$ to obtain the result, using the fact that

$$\begin{aligned} &\mathbb{P}\left((1-\epsilon)\|u-v\|^2 \le \frac{1}{n\sigma^2}\|Mu-Mv\|^2 \le (1+\epsilon)\|u-v\|^2\right) \\ &= 1 - \mathbb{P}\left(\frac{1}{n}\|Mu-Mv\|^2 < (1-\epsilon)\sigma^2\|u-v\|^2\right) \\ &\quad - \mathbb{P}\left(\frac{1}{n}\|Mu-Mv\|^2 > (1+\epsilon)\sigma^2\|u-v\|^2\right), \end{aligned} \tag{5.25}$$

and the choice $L^2 = 1/\sigma^2$. □

An easy way to guarantee that the multiplication of a vector with a matrix $y = Mx$ does not lose any information is to require that the matrix must be invertible, so that the initial vector x can be retrieved exactly from y. Of course, in our case, M has dimensions $n \times m$ with $n \gg m$, and thus cannot be invertible. However, we can still ensure that M has maximum rank m. In particular, we will show that with high probability any $m \times m$ submatrix of our random matrix M will be invertible. This is based on modifying a result from [156, Theorem 8.9].

Theorem 5.3.5 (Invertibility). *Given the matrix $M \in \mathbb{R}^{n\times m}$, where $n > m$, whose entries are each the difference of independent and identically distributed Bernoulli random variables with parameter $p \in (0,1)$, let $\mathcal{M} \in \mathbb{R}^{m\times m}$ be any square submatrix of M. Then, it holds for any $\epsilon > 0$ that*

$$\mathbb{P}(|\det(\mathcal{M})| \ge (2p(1-p))^{m/2}\sqrt{m!}\exp(-m^{1/2+\epsilon})) = 1 - o(1),$$

as $m \to \infty$. In particular, a submatrix $\mathcal{M}$ is invertible with probability at least $1 - o(1)$ as $m \to \infty$.

Proof. The matrix $R = \frac{1}{\sigma}\mathcal{M}$ has entries that are bounded and have mean zero and variance one. This means that it satisfies the hypotheses of [156, Theorem 8.9], so we can conclude that

$$\mathbb{P}(|\det(R)| \ge \sqrt{m!}\exp(-m^{1/2+\epsilon})) = 1 - o(1), \tag{5.26}$$

as $m \to \infty$. Using the fact that $\sigma = \sqrt{2p(1-p)}$ and $\det(R) = \sigma^{-m}\det(\mathcal{M})$ gives the result. $\square$

Theorem 5.3.5 says that the matrix $M \in \mathbb{R}^{n\times m}$ has submatrices that are likely to be invertible if the smaller dimension is sufficiently large. That is, they are likely to be invertible if the dimension of the feature vector that is the input to the transformation is sufficiently large. In practice, of course, any matrices will have finite size. In Figure 5.6 we calculate the probability (averaged over 10^4 independent realizations) of a given submatrix of M being invertible, for two different values of p. For both $p = 0.05$ and $p = 0.1$, we see that a submatrix is invertible at least 99 % of the time when the dimension $m = 100$ (in fact, when $p = 0.1$, the 99 % threshold is reached when $m = 48$). Since the dimension of the initial feature vector is likely to be large, any submatrices are highly likely to be invertible. Below, we will generally use $p = 0.05$ and $m = 433$, in which case the submatrices are almost guaranteed to be invertible (the probability of being singular is negligibly small).

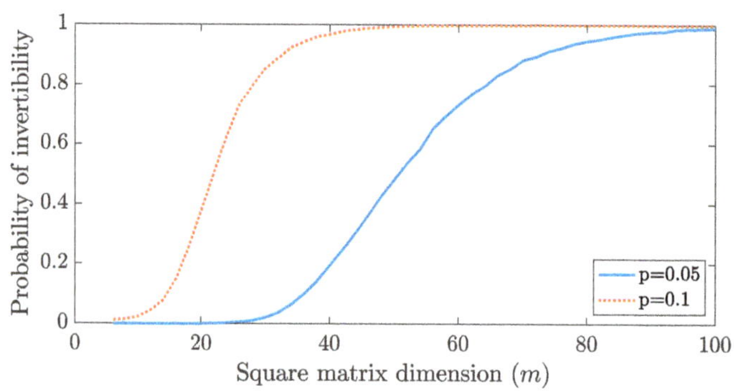

Figure 5.6: The probability of a submatrix of the random matrix M being invertible. When the dimension of the square submatrix is arbitrarily large, the probability that it is invertible approaches one. However, for finite matrix dimensions, very high probabilities of invertibility can be obtained with relatively small matrices.

The second part of the transformation A, defined in (5.15), is the application of a cap operator. This is a map $c_k : \mathbb{R}^n \to \mathbb{R}^n$ that retains the k largest (in magnitude) entries of a vector and sets all the others to zero. A cap operator is a crude way to sparsify a vector, with a degree of sparsity that can be controlled by varying the parameter k.

It is trivially the case that if $1 \le k \le n$, then for any $x \in \mathbb{R}^n$ and any $p \in (0,\infty)$, it holds that

$$\|x\|_\infty \le \|c_k(x)\|_p \le \|x\|_p. \tag{5.27}$$

In fact, if x is sufficiently sparse, in the sense that $\|x\|_p$ is small when p is small, then $c_k(x)$ is close to x. Various results along these lines exist, we prove one such statement below. A different version can be found in [70], for example.

Theorem 5.3.6. *Let $c_k : \mathbb{R}^n \to \mathbb{R}^n$ be the cap operation that retains the k largest (in magnitude) entries of a vector and sets all the others to zero. Then, for any $p \in (0, 2)$,*

$$\|x - c_k(x)\|_2 \le \|x\|_p (k+1)^{p+1} \quad \text{for all } x \in \mathbb{R}^n.$$

Proof. We show this by induction on k. First, consider the case where $k = 0$. By definition of the cap operation, $c_k(x)$ is then simply the zero vector, and we have $\|x\|_2 \le \|x\|_p$ for $0 < p < 2$ by monotonicity of the norms.

Suppose now that the property holds for $k-1$. Let $c_k^*(x)$ be the vector obtained from x by keeping its kth largest entry intact and setting all others to zero. We then have

$$\|x - c_k(x)\|_2 = \|x - c_{k-1}(x) - c_k^*(x)\|_2 \le \|x - c_{k-1}(x)\|_2 + \|c_k^*(x)\|_2.$$

From the inductive hypothesis, we have that

$$\|x - c_{k-1}(x)\|_2 \le \|x\|_p k^{p+1}. \tag{5.28}$$

Moreover, it holds that

$$\|c_k^*(x)\|_2 \le \|x\|_2 \le \|x\|_p \quad \text{for } p \in (0, 2). \tag{5.29}$$

Thus, we obtain

$$\|x - c_k(x)\|_2 \le \|x\|_p k^{p+1} + \|x\|_p \le \|x\|_p (k+1)^{p+1},$$

where the final inequality holds from the fact that the function $f(x) = (x+1)^{p+1} - x^{p+1} - 1$ satisfies $f'(x) = (p+1)[(x+1)^p - x^p] > 0$ for $x > 0$ and $f(0) = 0$, meaning that $f(x) > 0$ for all $x > 0$. The result then follows by induction. □

The consequence of these results is that, thanks to (5.27), the effect of the cap operation is always bounded (in the sense that $\|c_k\|_{\text{op}} < 1$) and, thanks to Theorem 5.3.6, if the initial feature vector has some sparsity, then this effect will in fact be correspondingly small (in the sense that $c_k(x)$ is close to x). Conversely, in the numerical experiments presented in the following section, we will show that the algorithm performs well on classification problems even when the data (and also the projected data) are not sparse. This shows that the random projections succeed in encoding the important information in a small number of the coefficients (as will be demonstrated by the fact that if the random projection step is removed, then the classification accuracy drops).

We would like to explore the extent to which the bio-inspired transformation (5.15), which yields sparse representations of signals through the use of random projections and a cap operation, can be used in classification problems. Recall that the main inspiration for this transformation was the function of the fruit fly's olfactory system, where the corresponding system's role is to facilitate the classification of odors. As a demonstrative classification problem, we chose to attempt musical genre classification. We used the GTZAN dataset [159] which consists of 30-second long extracts of music from 10 different genres. To generate the feature vector for our numerical experiments, we use the scattering transform followed by an average in time. This generated feature vectors with 433 entries. The scattering transform is a cascading sequence of alternating wavelet transforms and modulus operators that outputs coefficients that are locally invariant to translations and stable to deformations [23]. The software for the scattering transform can be found online at [24]. However, we could alternatively have used the subwavelength scattering transform (5.8), derived in Section 5.1.

To obtain the random matrix used for projection, we first sampled two random matrices whose entries were independent Bernoulli random variables with parameter p and then computed the difference of those two matrices. This gives the desired random matrix whose coefficients have mean zero, as described above. After random projection, we applied the cap operation to the feature vectors, retaining the k entries that were largest in absolute value and setting all others to zero. Given the resulting vector, classification was performed using a linear support vector machine.

We performed several experiments to understand the role of the parameters of the transformation A. In particular, n is the dimension of the space into which our random matrix projects the signal vectors. The parameter p corresponds to the Bernoulli parameter we use to sample the random matrix and, finally, the parameter k indicates how many coefficients we keep intact after the cap operation.

We initially performed classification on the feature vectors without the bio-inspired transformation, as a point of reference, to see if the random projection and cap operation improved or worsened the results. The resulting accuracy was consistently around 77 %. This is shown by the dotted lines in Figures 5.7, 5.8, and 5.9.

Our next experiment was to add a random projection and understand the effect of varying the distribution parameter p. We performed those experiments for both smaller

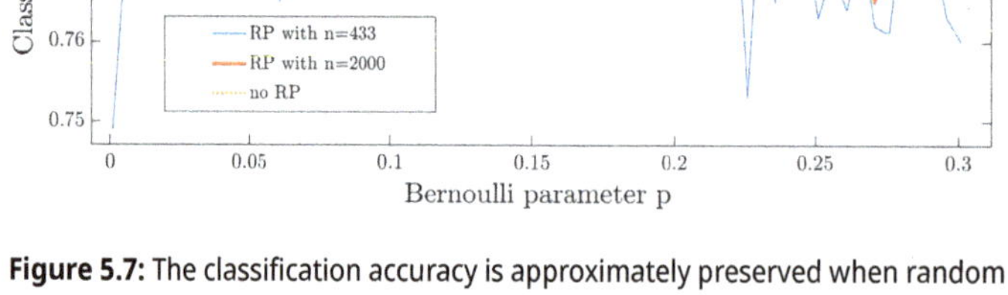

Figure 5.7: The classification accuracy is approximately preserved when random projections (RPs) are introduced and is not significantly affected by the Bernoulli parameter p. The classification accuracy is computed when using random projection of fixed dimension n, while varying the Bernoulli parameter and not using any cap operation.

($n = 433$) and larger ($n = 2000$) image spaces, to see if there was a clear advantage to projecting the feature vectors into a much higher dimensional space. The results can be seen in Figure 5.7, which suggest that there is no clear relation between the Bernoulli parameter p and classification accuracy. As shown above, the entries of our random matrix M are zero with probability $\mathbb{P}(m_{ij} = 0) = 2p^2 - 2p + 1$. This expression is strictly decreasing for $p < 0.5$, so the entries of M are less likely to be zero when p is larger. It is therefore in our interest to keep p small in subsequent experiments, so that our random matrix is likely to be sparser, which speeds up the calculations. In the following experiments, given that the effect on classification appears to be minimal, we fix the value $p = 0.05$.

As well as suggesting that the value of p does not greatly influence the classification accuracy, Figure 5.7 may suggest that applying a random projection slightly improves the classification accuracy. This is particularly the case for the larger dimension $n = 2000$, where the average success rate was more than a percentage point higher than for the classification without any random projection. However, this difference is not significant enough to be able to draw convincing conclusions at this point.

To evaluate the effect of the projection dimension n, we fixed the Bernoulli parameter at $p = 0.05$ and increased the dimension of the random matrix from $n = 433$ (the dimension of the feature vectors output by the scattering transform) up to $n = 2833$, adding 100 rows every time. The results can be seen in Figure 5.8. The accuracy is relatively stable across all values of n, and is similar to the performance of the original feature vectors without random projection. This result is maybe not surprising, as no additional information is being added by the random projection, instead it is merely being shuffled at random. It is, however, reassuring to observe that no information is lost or corrupted by adding a random projection. Once again, there appears to even be a

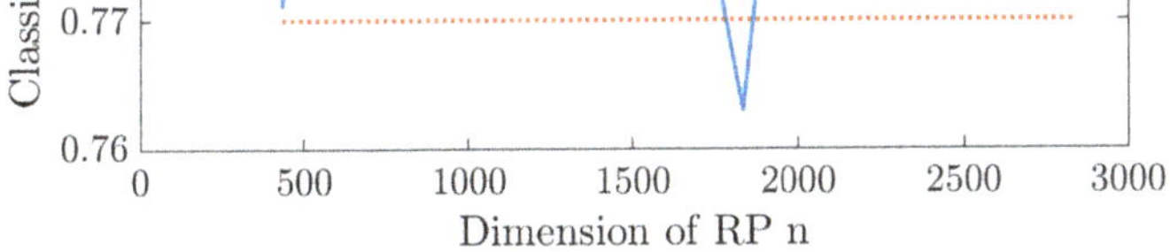

Figure 5.8: The classification accuracy is approximately preserved when random projections (RPs) are introduced and is not significantly affected by the dimension of the space into which we project. The classification accuracy is computed when using random projection of varying dimension n, fixed Bernoulli parameter $p = 0.05$, and no cap operation.

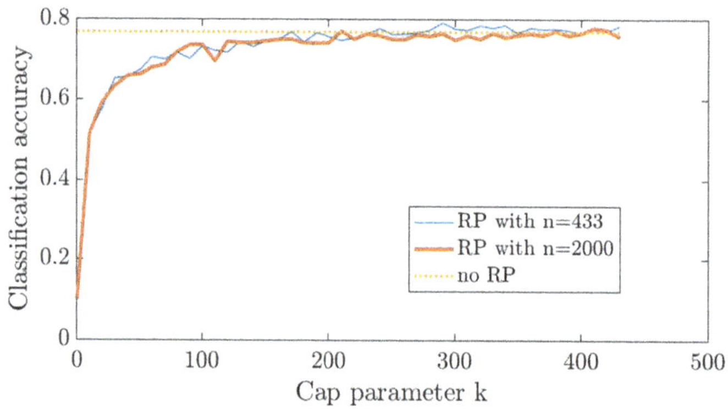

Figure 5.9: For sufficiently large cap parameter k, the introduction of a cap operation does not greatly influence the classification accuracy. This holds even when k is significantly smaller than the dimension of the initial feature vector (433). Classification accuracy is computed when using random projections (RPs) of fixed dimensions 433 and 2000, Bernoulli parameter $p = 0.05$, and a varying cap parameter k.

slight improvement in classification accuracy, thanks to the introduction of the random projection.

The effect of the cap operation on the classification accuracy can also be tested. We kept the Bernoulli parameter at $p = 0.05$ and added a cap operation with varying parameter k. We recorded classification accuracy for two different random projections; one with dimension $n = 433$ and the other with $n = 2000$. The results can be found in Figure 5.9. For very small values of k, the classification accuracy drops away quickly (down to a limiting value of 10 % when $k = 0$, in which case the ten classes are allocated without any retained information). However, the accuracy is rather stable for larger values of k. In particular, $k \approx 200$ is sufficient in both cases to attain a classification accuracy above 75 %. This result is noteworthy as $k = 200$ is less than half the size of the initial feature vectors outputted by the scattering transform. As previously with Bernoulli parameter p, it is in our interest to keep k small, so that the final feature vector is as sparse as possible. However, it is worth noting that our experiments revealed that increasing the cap parameter k did not dramatically decrease the training time of the support vector machine: with the Bernoulli parameter set at $p = 0.05$ and no cap operation the training time was 24.0 seconds, while setting a cap of $k = 200$ led to a training time of 23.3 seconds. This is because the linear support vector machine is not set up to be able to take advantage of the sparsity.

Taken together, these experiments show that projecting randomly into a space of seemingly arbitrary dimension with a matrix with small Bernoulli parameter before capping to leave just a couple of hundred entries gives an algorithm with good classification accuracy. In particular, we take projection into a space with dimension $n = 2000$, with Bernoulli parameter $p = 0.05$ and cap parameter $k = 200$ as our gold standard. Figure 5.10 shows the comparison of this transformation with the application of a cap alone.

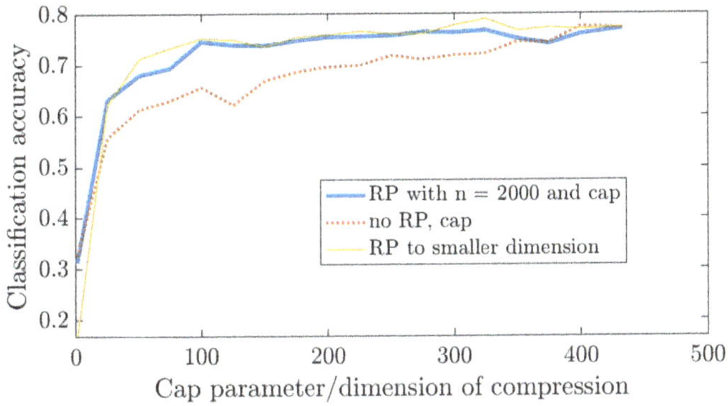

Figure 5.10: The application of a random projection (RP) together with a cap consistently shows higher accuracy than the use of a cap alone. On the other hand, a simple random projection to a space of smaller dimension performs similarly to the random projection and cap. All three methods show relatively good classification accuracy, as long as the cap/dimension of compression is above 100. Both random projections in this figure were performed with a Bernoulli parameter $p = 0.05$.

The full transformation (with random projection followed by the cap) consistently performs better than the cap alone, showing that the application of a random projection is important for retaining information when the vectors are truncated using the cap operator. We also compared the performance of our transformation A with a simple random compression. This random compression was performed by multiplying by the same random matrix used in our bioinspired transformation, with a Bernoulli parameter $p = 0.05$, but with the dimension decreased. The image dimension of the random compression is shown on the horizontal axis of Figure 5.10, so that it can be compared to the cap parameter in the two other algorithms. All three sets of results show that the accuracy does not drop significantly as long as the number of coefficients retained is above 100, thus suggesting that we can easily decrease the dimension of the feature vectors (which initially have 433 entries), thus making them sparser, without impeding our ability to classify them.

Our final experiment focused on determining whether the addition of the bioinspired transformation A could improve the robustness of the classification. We added random Gaussian noise to the feature vectors outputted by the scattering transform and compared three cases: classifying the noisy vectors directly without any transformation, classifying the noisy vectors that have been randomly projected to a space of dimension $n = 2000$ and, finally, classifying the noisy vectors after applying both the random projection and a cap operation with a parameter $k = 200$. The results for these three cases are reported in Figure 5.11. In each case the noise was generated independently with zero mean and increasing standard deviation. It is clear from Figure 5.11 that adding the random projection increases the robustness of the classification. Classification with the projected vectors consistently performs a few percent better than without any random projection. Conversely, the addition of a cap operation seems to

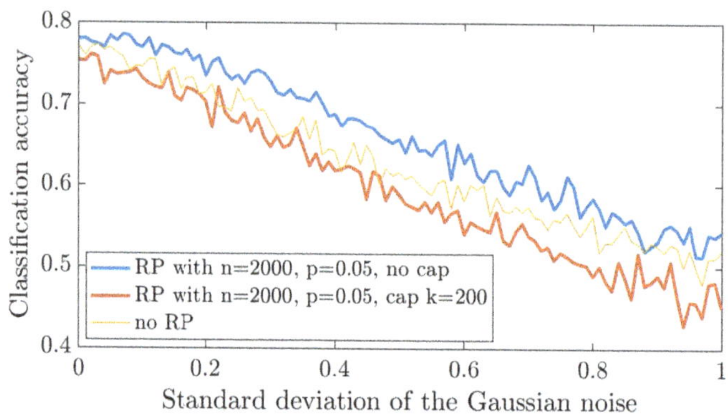

Figure 5.11: The classification accuracy shows improved robustness to noise when random projections (RPs) are introduced. This effect is, however, lost when a cap operator is added. Classification accuracy is computed when adding Gaussian noise of mean zero and varying standard deviation to the initial feature vectors.

yield slightly worse results, suggesting that there is a trade-off between robustness and sparsity.

These results demonstrate yet another successful application of biomimicry, this time to a classification problem. Our two-step signal transform, inspired by the function of the fruit fly's olfactory system, has the ability to sparsify the data while preserving the classification accuracy. Our experiments showed that it also leads to robustness benefits, giving improved classification accuracy when random errors are added to the data. Perhaps most importantly, the signal transform is very simple and requires very little computational power to execute, giving a distinct advantage over more intricate or learning-based approaches.

5.4 Discussion

We began by studying an array of subwavelength resonators that has similar dimensions to the cochlea and mimics its biomechanical properties. We used concise asymptotic results, which were derived from first principles, to allow us to efficiently fine-tune the design of our structure to mimic the cochlea. Using this analysis, we derived the corresponding signal processing algorithm, known here as the subwavelength scattering transform. The use of a cochlea-like structure as an intermediate step overcame the challenges posed by modeling the biological cochlea directly. We were then able to build upon this by adding additional processing steps tailored to the class of natural sounds, directly inferred from the performance of the human auditory system and other biological sensing systems.

The work presented in this chapter contributes to developing the precise mathematical foundations for the exchange of design principles and features between biological auditory systems, artificial sound-filtering devices and signal processing algorithms, as depicted in Figure 1.2. Biomimicry has already had a significant impact on both the development of artificial hearing approaches and, conversely, on our understanding of biological auditory systems. The mathematical foundations to support the development of this powerful methodology can play an important role in answering many of the open questions that concern how we interact with sound waves in our environment.

The results in this chapter showed the potential benefits of taking inspiration from biology. Sparsity and robustness are important properties for biological sensing systems. Animals have limited neural bandwidth so need to be able to encode information as efficiently as possible. Sparsifying data is one way of achieving this, as we did in Section 5.3. Another is using known statistical properties of target data sets to obtain low-dimensional representations, for example, by using the natural sound coefficients we considered in Section 5.2. Other strategies include using compressive non-linearities to rescale data. Partly for this reason, we will consider *active* cochlea-inspired metamaterials in Chapter 7. Robustness is similarly important for an animal's ability to understand its noisy environment and there are many examples of biological systems demonstrating remarkable robustness properties. We will explore this in more detail in Chapter 6.

6 Robustness with respect to imperfections

A significant consideration when designing metamaterials and other devices that include structures at subwavelength (sometimes microscopic) scales is whether their function is robust. When any such metamaterial or device is manufactured, it will inevitably include some small deviations and defects. On top of this, a system is liable to pick up further imperfections and damage during its usage. For devices to be commercially viable, it is helpful if they do not fail in response to the smallest damage.

Biological systems face a similar challenge. Cell development is a complex and sometimes unpredictable process and systems often need to continue functioning even after years of exposure to harsh conditions (depending on the organism's habitat and lifestyle choices). As a result, it is a significant advantage for an organism to have sensing systems that are robust with respect to damage.

Fortunately, the biological cochlea has a remarkable ability to function effectively even when significantly damaged. As depicted in Figure 6.1, cochlear receptor cells are often significantly damaged in older organisms, particularly if they have been exposed to loud sounds for long periods. However, it has been observed that humans can lose as much as 30–50 % of their receptor cells without any perceptible loss of hearing function [48, 169] (see Figure 6.1). This remarkable robustness poses a timely question for us: how do cochlea-inspired rainbow sensors behave under similar errors and imperfections? This chapter will use our asymptotic formulas to give quantitative insight into the answers to this question.

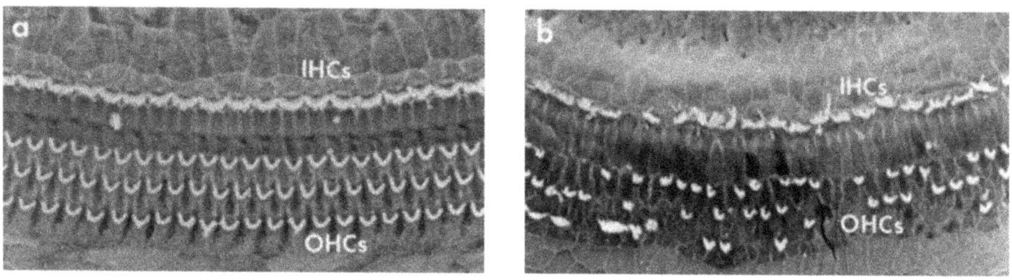

Figure 6.1: The receptor cells in a (a) normal and (b) damaged cochlea. The receptor cells are arranged as one row of inner hair cells (IHCs) and three rows of outer hair cells (OHCs). In a damaged cochlea, the stereocilia are severely deformed and, in many cases, missing completely. The images are scanning electron micrographs of rat cochleae, provided by Elizabeth M Keithley.

6.1 Symmetric matrices and diluteness

We will consider the same three-dimensional high-contrast metamaterial as in Chapter 4. Recall, in particular, that we are able to characterize the resonant frequencies and associated resonant modes in terms of the eigenstates of the generalized capacitance

https://doi.org/10.1515/9783110784961-006

matrix. This approach is both computationally efficient and also allows us to make analytic statements, as we will do below.

For the analysis in this chapter, we will want to take advantage of some of the existing theory for describing matrix eigenvalue perturbations. For technical reasons, much of this theory has been developed for the specific case of symmetric matrics. Recall that in (4.13) we defined the generalized capacitance matrix as

$$\mathcal{C}_{ij} := \frac{1}{|D_i|} C_{ij}.$$

In this chapter, we will instead work with a symmetric version of the generalized capacitance matrix, which we will denote by $\mathcal{C}^{\mathrm{s}}$. If we define the volume scaling matrix $V \in \mathbb{R}^{N\times N}$ to be the diagonal matrix given by

$$V_{ii} = \frac{1}{\sqrt{|D_i|}}, \qquad i = 1, \dots, N, \tag{6.1}$$

then the standard generalized capacitance matrix is given by $\mathcal{C} = V^2 C$. In this chapter, we will work with the *symmetric* generalized capacitance matrix $\mathcal{C}^{\mathrm{s}} \in \mathbb{R}^{N\times N}$, defined as

$$\mathcal{C}^{\mathrm{s}} = VCV.$$

Importantly, $\mathcal{C}^{\mathrm{s}} = VCV$ is similar to $\mathcal{C} = V^2 C$, meaning they have the same eigenvalues (with the same multiplicity). As a result, we have an analogous result to Theorem 4.2.1 describing the resonant frequencies and an associated formula for the resonant modes.

Theorem 6.1.1. *Consider a system of N subwavelength resonators in* $\mathbb{R}^3$ *and let* $\{(\lambda_n, \underline{v}_n) : n = 1, \dots, N\}$ *be the eigenpairs of the (symmetric) generalized capacitance matrix* $\mathcal{C}^{\mathrm{s}} \in \mathbb{R}^{N\times N}$. *As* $\delta \to 0$, *the subwavelength resonant frequencies satisfy the asymptotic formula*

$$\omega_n^{\pm} = \pm\sqrt{v_b^2 \lambda_n \delta} - \mathrm{i}\tau_n \delta + O(\delta^{3/2}),$$

for $n = 1, \dots, N$, *where the second-order coefficients* τ_n *are given by*

$$\tau_n = \frac{v_b^2}{8\pi v} \frac{1}{\|\underline{v}_n\|^2} \underline{v}_n^{\top} VCJCV \underline{v}_n, \qquad n = 1, \dots, N,$$

with J being the $N \times N$ *matrix of ones and* $\|\cdot\|$ *the standard* l^2 *norm.*

Corollary 6.1.1. *Let* $\underline{v}_n$ *be the normalized eigenvector of* $\mathcal{C}^{\mathrm{s}}$ *associated to the eigenvalue* λ_n. *Then the normalized resonant mode* u_n *associated to the resonant frequency* ω_n *is given, as* $\delta \to 0$, *by*

$$u_n(x) = \begin{cases} \underline{v}_n^{\top} V \mathbf{S}_D^{k_0}(x) + O(\delta^{1/2}), & x \in \mathbb{R}^3 \setminus \overline{D}, \\ \underline{v}_n^{\top} V \mathbf{S}_D^{k}(x) + O(\delta^{1/2}), & x \in D, \end{cases}$$

where $\mathbf{S}_D^k : \mathbb{R}^3 \to \mathbb{C}^N$ *is the vector-valued function given by*

$$\mathbf{S}_D^k(x) = \begin{pmatrix} \mathcal{S}_D^k[\psi_1](x) \\ \vdots \\ \mathcal{S}_D^k[\psi_N](x) \end{pmatrix}, \quad x \in \mathbb{R}^3 \setminus \partial D,$$

with $\psi_i := (\mathcal{S}_D^0)^{-1}[\chi_{\partial D_i}]$.

In this case, since C is symmetric, V is diagonal and J is positive semidefinite, it is easy to see that $\tau_n \geq 0$ for all $n = 1, \dots, N$.

We will shortly want to study how the properties of the symmetric generalized capacitance matrix $\mathcal{C}^s$ vary when changes are made to the structure D. For this reason, we will often write $\mathcal{C}^s = \mathcal{C}^s(D)$ to emphasize the dependence of the generalized capacitance matrix on the geometry of D. Similarly, we will write $\lambda_i = \lambda_i(D)$ and $\tau_i = \tau_i(D)$ for the quantities from Theorem 6.1.1. With this in mind, it is important to notice that the asymptotic expansion in Theorem 6.1.1 is uniform with respect to geometric perturbations that keep the resonators separated (this breaks down if they touch or overlap). This is a useful property of this result which has been used in many places, such as in [9, Theorem 2], where the result for ϵ-small resonators is proved as a modification of [11, Theorem 3.2.5].

We will begin by deriving formulas to describe the effects of making small perturbations to the positions and sizes of the resonators, as depicted in Figure 6.2. Perturbations of this nature are important as they are likely to be introduced when a device is manufactured. The results in this section give quantitative estimates on the extent to which the perturbations of the structure's properties are stable with respect to small imperfections.

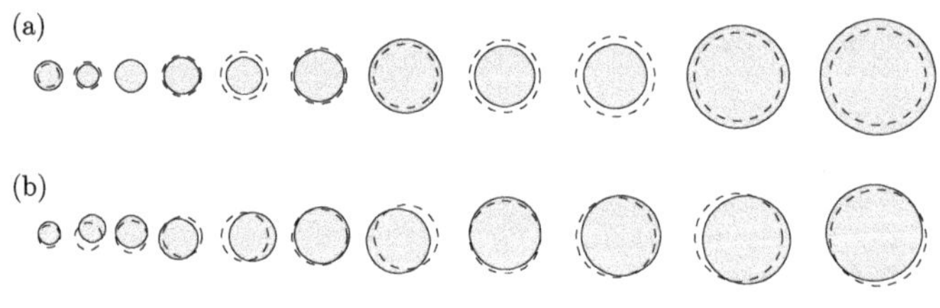

Figure 6.2: We study the effects of adding random perturbations to the (a) size and (b) position of the resonators in a cochlea-inspired rainbow sensor. The original structure is shown in dashes.

In order to simplify the analysis, and to allow us to work with explicit formulas, we will make an assumption that the resonators are small compared to the distance between them. In particular, we will assume that each resonator D_i is given by $B_i + \epsilon^{-1} z_i$ where $B_i \subset \mathbb{R}^3$ is some fixed domain, $z_i \in \mathbb{R}^3$ is some fixed vector and $0 < \epsilon \ll 1$ is some small parameter. We will assume that each fixed domain B_i, for $i = 1, \dots, N$, is positioned so that it contains the origin and that the complete structure is given by

$$D = \bigcup_{i=1}^{N} D_i, \quad D_i = (B_i + \epsilon^{-1} z_i). \tag{6.2}$$

Under this assumption, the generalized capacitance matrix has an explicit leading-order asymptotic expression in terms of the *dilute* generalized capacitance matrix:

Definition 6.1.1 (Dilute generalized capacitance matrix). Given $0 < \epsilon \ll 1$ and a resonator array that is ϵ-dilute in the sense of (6.2), the associated dilute generalized capacitance matrix $\mathcal{C}^\epsilon \in \mathbb{R}^{N \times N}$ is defined as

$$\mathcal{C}_{ij}^\epsilon = \begin{cases} \frac{\mathrm{Cap}_{B_i}}{|B_i|}, & i = j, \\ -\epsilon \frac{\mathrm{Cap}_{B_i} \mathrm{Cap}_{B_j}}{4\pi |z_i - z_j| \sqrt{|B_i||B_j|}}, & i \neq j, \end{cases}$$

where we define the capacitance Cap_B of a set $B \subset \mathbb{R}^3$ to be the strictly positive number given by

$$\mathrm{Cap}_B := -\int_{\partial B} (\mathcal{S}_B^0)^{-1} [\chi_{\partial B}] \, \mathrm{d}\sigma.$$

Lemma 6.1.1. *Consider a resonator array that is ϵ-dilute in the sense of* (6.2). *In the limit as $\epsilon \to 0$, the asymptotic behavior of the (symmetric) generalized capacitance matrix is given by*

$$\mathcal{C}^s = \mathcal{C}^\epsilon + O(\epsilon^2) \quad \text{as } \epsilon \to 0.$$

Proof. This was proved in [9, Lemma 1] as a modification of the original result in [12, Lemma 4.3]. □

It would also be possible to state an appropriate diluteness condition as a rescaling of the sizes of the resonators, by taking $D_i = \epsilon B_i + z_i$ in (6.2). This would give analogous but rescaled results, as used for the analysis in [12].

6.2 Imperfections in the device

In Figure 6.3 we show how the subwavelength resonant frequencies of a system of 22 resonators changes as random errors are added to the size and position of the resonators. In both cases, it is clear that the perturbations of the spectrum are continuous in the sense that the original spectrum is recovered as the magnitude of the perturbations goes to zero. We wish to characterise this behaviour in this section. We first consider imperfections due to changes in the size of the resonators. In particular, suppose there exist some factors $\alpha_1, \dots, \alpha_N$ such that the perturbed structure is given by

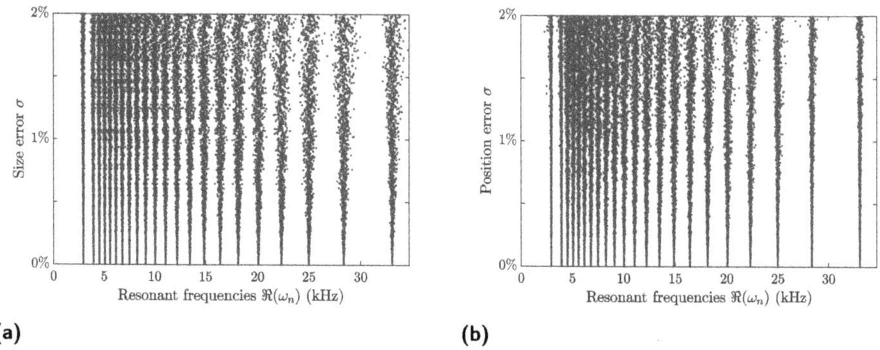

Figure 6.3: The effect of random errors and imperfections on the subwavelength resonant frequencies of a cochlea-inspired rainbow sensor. (a) Random errors are added to the sizes of the resonators. (b) Random errors are added to the positions of the resonators. In both cases the errors are Gaussian with mean zero and variance σ^2. These simulations are performed on the full differential problem using the multipole expansion method. The deviation of the random error σ is expressed as a percentage of the unperturbed values.

$$D^{(\alpha)} = \bigcup_{i=1}^{N} ((1+\alpha_i)B_i + \epsilon^{-1} z_i). \tag{6.3}$$

We will assume that the perturbations $\alpha_1, \dots, \alpha_N$ are small in the sense that there exists some parameter α such that $\alpha_i = O(\alpha)$ as $\alpha \to 0$.

Lemma 6.2.1. *Suppose that a resonator array D is deformed to give $D^{(\alpha)}$, as defined in (6.3), and that the size change parameters $\alpha_1, \dots, \alpha_N$ satisfy $\alpha_i = O(\alpha)$ as $\alpha \to 0$ for all $i = 1, \dots, N$. Then, for fixed $0 < \epsilon \ll 1$, the dilute generalized capacitance matrix associated to $D^{(\alpha)}$ is given by*

$$\mathcal{C}^\epsilon(D^{(\alpha)}) = \mathcal{C}^\epsilon(D) + A(\alpha),$$

where $A(\alpha)$ is a symmetric $N \times N$-matrix whose Frobenius norm satisfies $\|A\|_F = O(\alpha)$ as $\alpha \to 0$. Furthermore, the error bound $\|\mathcal{C}^\epsilon(D^{(\alpha)}) - \mathcal{C}^\epsilon(D)\|_F = O(\alpha)$ as $\alpha \to 0$ is uniform with respect to $\epsilon \in [0,1]$.

Proof. Making the substitution $B_i \mapsto (1+\alpha_i)B_i$ in Definition 6.1.1 gives

$$\mathcal{C}^\epsilon_{ij}(D^{(\alpha)}) = \begin{cases} \dfrac{\mathrm{Cap}_{B_i}}{(1+\alpha_i)|B_i|}, & i = j, \\ -\epsilon \dfrac{\mathrm{Cap}_{B_i}\mathrm{Cap}_{B_j}}{4\pi|z_i - z_j|\sqrt{(1+\alpha_i)(1+\alpha_j)}\sqrt{|B_i||B_j|}}, & i \neq j. \end{cases}$$

For small α we can expand the denominators to obtain

$$\mathcal{C}^\epsilon_{ij}(D^{(\alpha)}) = \begin{cases} (1-\alpha_i)\dfrac{\mathrm{Cap}_{B_i}}{|B_i|} + O(\alpha^2), & i = j, \\ -\epsilon\Big[(1 - \frac{1}{2}(\alpha_i + \alpha_j))\dfrac{\mathrm{Cap}_{B_i}\mathrm{Cap}_{B_j}}{4\pi|z_i - z_j|\sqrt{|B_i||B_j|}} + O(\alpha^2)\Big], & i \neq j, \end{cases}$$

as $\alpha \to 0$. From this, we can see that

$$A_{ij} = C^{\epsilon}_{ij}(D^{(\alpha)}) - C^{\epsilon}_{ij}(D) = \begin{cases} -\alpha_i \frac{\mathrm{Cap}_{B_i}}{|B_i|} + O(\alpha^2), & i = j, \\ \epsilon\left[\frac{1}{2}(\alpha_i + \alpha_j)\frac{\mathrm{Cap}_{B_i}\mathrm{Cap}_{B_j}}{4\pi|z_i - z_j|\sqrt{|B_i||B_j|}} + O(\alpha^2)\right], & i \neq j, \end{cases}$$

as $\alpha \to 0$. To see that the convergence of $C^{\epsilon}(D^{(\alpha)})$ to $C^{\epsilon}(D)$ is uniform in $\epsilon \in [0,1]$, notice that the diagonal terms of A do not depend on ϵ and the absolute value of the off-diagonal terms is a monotonic function of ϵ. □

Theorem 6.2.1. *Suppose that a resonator array D is ϵ-dilute in the sense of* (6.2) *and is deformed to give $D^{(\alpha)}$, as defined in* (6.3), *for size change parameters $\alpha_1, \dots, \alpha_N$ which satisfy $\alpha_i = O(\alpha)$ as $\alpha \to 0$ for all $i = 1, \dots, N$. Then, the resonant frequencies satisfy*

$$|\omega_n(D) - \omega_n(D^{(\alpha)})| = O(\sqrt{\delta(\alpha + \epsilon^2)}),$$

as $\alpha, \delta, \epsilon \to 0$.

Proof. From Lemma 6.2.1 we have that $C^{\epsilon}(D^{(\alpha)}) = C^{\epsilon}(D) + A(\alpha)$, where A is a symmetric $N \times N$-matrix. Then, by the Wielandt–Hoffman theorem [85], it holds that the eigenvalues of $C^{\epsilon}(D)$ and $C^{\epsilon}(D^{(\alpha)})$, which we denote by $\lambda^{\epsilon}_i(D)$ and $\lambda^{\epsilon}_i(D^{(\alpha)})$, respectively, satisfy

$$\sum_{n=1}^{N} \left(\lambda^{\epsilon}_n(D) - \lambda^{\epsilon}_n(D^{(\alpha)})\right)^2 \leq \|A\|_F^2. \tag{6.4}$$

From this we can see that $|\lambda^{\epsilon}_i(D) - \lambda^{\epsilon}_i(D^{(\alpha)})| = O(\alpha)$ as $\alpha \to 0$, since $\|A\|_F = O(\alpha)$ as $\alpha \to 0$ by Lemma 6.2.1. Further, this convergence is uniform in ϵ and δ (from Lemma 6.2.1 and since these quantities do not depend on δ). By a similar argument, and using Lemma 6.1.1, we have that

$$|\lambda_n(D) - \lambda^{\epsilon}_n(D)| = O(\epsilon^2) \quad \text{and} \quad |\lambda_n(D^{(\alpha)}) - \lambda^{\epsilon}_n(D^{(\alpha)})| = O(\epsilon^2), \quad \text{as } \epsilon \to 0. \tag{6.5}$$

Again, we have that this convergence is uniform with respect to α and δ, since there is no dependence on either α or δ (crucially, $\lambda_n(D^{(\alpha)}) - \lambda^{\epsilon}_n(D^{(\alpha)})$ is constant as a function of α). Finally, we use Theorem 4.2.1 to find the resonant frequencies when $\delta \to 0$:

$$\begin{aligned} |\omega_n(D) - \omega_n(D^{(\alpha)})| &= \left|\sqrt{\delta v_b^2 \lambda_n(D)} - \sqrt{\delta v_b^2 \lambda_n(D^{(\alpha)})}\right| + O(\delta) \\ &\leq \sqrt{\delta v_b^2}\sqrt{|\lambda_n(D) - \lambda_n(D^{(\alpha)})|} + O(\delta). \\ &\leq \sqrt{\delta v_b^2}\sqrt{|\lambda_n(D) - \lambda^{\epsilon}_n(D)| + |\lambda^{\epsilon}_n(D) - \lambda^{\epsilon}_n(D^{(\alpha)})| + |\lambda^{\epsilon}_n(D^{(\alpha)}) - \lambda_n(D^{(\alpha)})|} + O(\delta). \end{aligned}$$

Combining this with (6.4) and (6.5) gives the result, provided that the $O(\delta)$ remainder term is well behaved as $\alpha, \epsilon \to 0$. Uniformity with respect to ϵ follows from [9, Theorem 2], and uniformity with respect to small values of α follows similarly. The crucial property is that Theorem 4.2.1 gives an expansion of this form for any configuration of nonoverlapping resonators. This is based on the asymptotic expansion (4.7) of $\mathcal{S}_D^k[\phi]$ as $k \to 0$, in which each term has the form $\mathcal{S}_{D,n}[\phi](x)$, for $n = 1, 2, \dots$, where

$$\mathcal{S}_{D,n}[\phi](x) = -\frac{\mathrm{i}^n}{4\pi n!} \int_{\partial D} |x - y|^{n-1} \phi(y)\, \mathrm{d}\sigma(y), \quad x \in \partial D. \tag{6.6}$$

The leading-order equation gives us that $\phi \in \operatorname{span}\{\psi_1, \dots, \psi_N\}$, where $\psi_i = (\mathcal{S}_D^0)^{-1}[\chi_{\partial D_i}]$. If we rescale one of the domains $D_i \mapsto (1 + \alpha_i)D_i$, then the quantities $\mathcal{S}_{D,n}[\psi_i](x)$ depend continuously on α_i. Thus, if α is sufficiently small that the resonators do not overlap, then taking the supremum of these continuous quantities over $\alpha_i \in (-\alpha, \alpha)$ gives a bound that holds uniformly over all such (sufficiently small) values of α. □

While the Wielandt–Hoffman theorem was used in (6.4), there are a range of results that could be invoked here. For example, if $\lambda_{\min}$ and $\lambda_{\max}$ are the smallest and largest eigenvalues of A then it holds that

$$\lambda_n^\epsilon(D) + \lambda_{\min} \le \lambda_i^\epsilon(D^{(\alpha)}) \le \lambda_n^\epsilon(D) + \lambda_{\max},$$

for all $n = 1, \dots, N$. For a selection of results on perturbations of eigenvalues of symmetric matrices, see [85].

Let us now consider imperfections due to changes in the positions of the resonators. In particular, suppose there exist some vectors $\beta_1, \dots, \beta_N \in \mathbb{R}^3$ such that the perturbed structure is given by

$$D^{(\beta)} = \bigcup_{i=1}^{N} (B_i + \epsilon^{-1}(z_i + \beta_i)). \tag{6.7}$$

We will assume that the perturbations $\beta_1, \dots, \beta_N$ are small in the sense that there exists some parameter $\beta \in \mathbb{R}$ such that $\|\beta_i\| = O(\beta)$ as $\beta \to 0$. We will proceed as in the previous section, by considering the dilute generalized capacitance matrix $\mathcal{C}^\epsilon$.

Lemma 6.2.2. *Suppose that a resonator array D is deformed to give $D^{(\beta)}$, as defined in (6.7), and that the translation vectors $\beta_1, \dots, \beta_N$ satisfy $\|\beta_i\| = O(\beta)$ as $\beta \to 0$ for all $i = 1, \dots, N$. Then, for fixed $0 < \epsilon \ll 1$, the dilute generalized capacitance matrix associated to $D^{(\beta)}$ is given by*

$$\mathcal{C}^\epsilon(D^{(\beta)}) = \mathcal{C}^\epsilon(D) + B(\beta),$$

where $B(\beta)$ is a symmetric $N \times N$-matrix whose Frobenius norm satisfies $\|B\|_F = O(\beta)$ as $\beta \to 0$. Furthermore, the error bound $\|\mathcal{C}^\epsilon(D^{(\beta)}) - \mathcal{C}^\epsilon(D)\|_F = O(\beta)$ as $\beta \to 0$ is uniform with respect to $\epsilon \in [0, 1]$.

Proof. We will make the substitution $z_i \mapsto z_i + \beta_i$ in Definition 6.1.1. The diagonal entries of C^ϵ are unchanged. For the off-diagonal entries, we have that

$$C_{ij}^\epsilon(D^{(\beta)}) = -\epsilon \frac{\mathrm{Cap}_{B_i}\mathrm{Cap}_{B_j}}{4\pi|z_i + \beta_i - z_j - \beta_j|\sqrt{|B_i||B_j|}}, \quad i \neq j.$$

For small β, we can expand the denominator to give

$$\frac{1}{|z_i + \beta_i - z_j - \beta_j|} = \frac{1}{|z_i - z_j|} - (\beta_i - \beta_j) \cdot \frac{z_i - z_j}{|z_i - z_j|^3} + O(\beta^2), \quad i \neq j,$$

as $\beta \to 0$. This gives us that

$$C_{ij}^\epsilon(D^{(\beta)}) = C_{ij}^\epsilon(D) + \epsilon(\beta_i - \beta_j) \cdot \frac{(z_i - z_j)\mathrm{Cap}_{B_i}\mathrm{Cap}_{B_j}}{4\pi|z_i - z_j|^3\sqrt{|B_i||B_j|}} + O(\beta^2), \quad i \neq j, \tag{6.8}$$

as $\beta \to 0$. The uniformity follows by taking the supremum of (6.8) with respect to $\epsilon \in [0,1]$. □

Theorem 6.2.2. *Suppose that a resonator array D is ϵ-dilute in the sense of* (6.2) *and is deformed to give $D^{(\beta)}$, as defined in* (6.7), *for translation vectors $\beta_1, \dots, \beta_N$ which satisfy $\|\beta_i\| = O(\beta)$ as $\beta \to 0$ for all $i = 1, \dots, N$. Then the resonant frequencies satisfy*

$$|\omega_n(D) - \omega_n(D^{(\beta)})| = O(\sqrt{\delta(\beta + \epsilon^2)}),$$

as $\beta, \delta, \epsilon \to 0$.

Proof. From Lemma 6.2.2, we have that $C^\epsilon(D^{(\beta)}) = C^\epsilon(D) + B(\beta)$ where B is a symmetric $N \times N$-matrix so we can proceed as in Theorem 6.2.1 to use the Wielandt–Hoffman theorem to bound $|\lambda_n^\epsilon(D) - \lambda_n^\epsilon(D^{(\beta)})|$ by $\|B\|_F$ for each $n = 1, \dots, N$. Then, approximating under the assumption that δ and ϵ are small gives the result. □

Recall the expansion $\omega_n = \sqrt{\delta v^2 \lambda_n} - \mathrm{i}\delta\tau_n + \cdots$ from Theorem 4.2.1. The formula for τ_n involves the eigenvectors $\underline{v}_n$ of the generalized capacitance matrix. Assuming that the material parameters are real, τ_n describes the leading-order imaginary part of the resonant frequency, so it is important to understand how it is affected by imperfections in the structure.

If we consider a resonator array D that is such that the associated (symmetric) generalized capacitance matrix $C(D)$ has N distinct, simple eigenvalues, then we can derive an approximate formula for the effects of perturbations on the eigenvectors of $C^s(D)$. Suppose that a perturbation, governed by the parameter γ, is made to the structure to give $D^{(\gamma)}$ and that there is a symmetric matrix $\Gamma(\gamma)$ which is such that

$$C^s(D^{(\gamma)}) = C^s(D) + \Gamma(\gamma), \tag{6.9}$$

where $\|\Gamma(\gamma)\| \to 0$ as $\gamma \to 0$. In this setting, we can derive an approximate formula for the perturbed eigenvector $\underline{v}_n(D^{(\gamma)})$.

Since $\mathcal{C}^s(D)$ is a symmetric matrix, it has an orthonormal basis of eigenvectors $\{\underline{v}_n : n = 1, \ldots, N\}$ with associated eigenvalues $\sigma(\mathcal{C}^s(D)) = \{\lambda_n : n = 1, \ldots, N\}$, which are assumed to be distinct. Under this assumption, we have the decomposition

$$(\lambda I - \mathcal{C}^s(D))^{-1} x = \sum_{k=1}^{N} \frac{\langle x, \underline{v}_k \rangle}{\lambda - \lambda_k} \underline{v}_k, \quad x \in \mathbb{C}^n, \lambda \in \mathbb{C} \setminus \sigma(\mathcal{C}^s). \tag{6.10}$$

From this we can see that $\|(\lambda I - \mathcal{C}^s(D))^{-1}\| \leq \operatorname{dist}(\lambda, \sigma(\mathcal{C}^s(D)))^{-1}$. If we add a perturbation matrix $\Gamma(\gamma)$ which is such that $\|\Gamma(\gamma)\| < \operatorname{dist}(\lambda, \sigma(\mathcal{C}^s(D)))$, then $\lambda I - \mathcal{C}^s(D^{(\gamma)}) = \lambda I - \mathcal{C}^s(D) - \Gamma(\gamma)$ is invertible. Further, in this case, we can use a Neumann series to see that

$$\begin{aligned}(\lambda I - \mathcal{C}^s(D^{(\gamma)}))^{-1} &= (\lambda I - \mathcal{C}^s(D) - \Gamma)^{-1} \\ &= (\lambda I - \mathcal{C}^s(D))^{-1} \sum_{i=0}^{\infty} \Gamma^i ((\lambda I - \mathcal{C}^s(D))^{-1})^i.\end{aligned}$$

Substituting the decomposition (6.10) and taking only the first two terms from this Neumann series expansion, we see that for a fixed $\lambda \in \mathbb{C} \setminus \sigma(\mathcal{C}^s)$ we have

$$(\lambda I - \mathcal{C}^s(D^{(\gamma)}))^{-1} = \sum_{k=1}^{N} \frac{\langle \cdot, \underline{v}_k \rangle}{\lambda - \lambda_k} \underline{v}_k + \sum_{k=1}^{N} \sum_{j=1}^{N} \frac{\langle \cdot, \underline{v}_j \rangle \langle \Gamma \underline{v}_j, \underline{v}_k \rangle}{(\lambda - \lambda_k)(\lambda - \lambda_j)} \underline{v}_k + \cdots, \tag{6.11}$$

where the remainder terms are $O(\|\Gamma(\gamma)\|^2)$ as $\gamma \to 0$.

Suppose we have a collection of closed curves $\{\eta_n : n = 1, \ldots, N\}$ which do not intersect and are such that the interior of each curve η_n contains exactly one eigenvalue λ_n. We know that we may choose γ to be sufficiently small that the eigenvalues of $\mathcal{C}^s(D^{(\gamma)})$ remain within the interior of these same curves. Thus, the operator $P_n : \mathbb{C}^N \to \mathbb{C}^N$, defined by

$$\mathcal{P}_n = \frac{1}{2\pi \mathrm{i}} \int_{\eta_n} (\lambda I - \mathcal{C}^s(D^{(\gamma)}))^{-1} \, \mathrm{d}\lambda, \tag{6.12}$$

is the projection onto the eigenspace associated to the perturbed eigenvalue $\lambda_n(D^{(\gamma)})$. Using the expansion (6.11), we can calculate an approximation to the operator $\mathcal{P}_n$, given by

$$\mathcal{P}_n \approx \langle \cdot, \underline{v}_n \rangle \underline{v}_n + \sum_{\substack{k=1 \\ k \neq n}}^{N} \frac{\langle \cdot, \underline{v}_n \rangle \langle \Gamma \underline{v}_n, \underline{v}_k \rangle}{(\lambda_n - \lambda_k)} \underline{v}_k,$$

where we are assuming the remainder term to be small in order for the approximation to hold. This is a technical issue, which is not trivial to show precisely due to the nonuniformity of the expansion (6.11) with respect to λ, particularly near to $\lambda \in \sigma(\mathcal{C}^s(D))$. Applying this approximation for the operator $\mathcal{P}_n$ to the unperturbed eigenvector $\underline{v}_n$ gives the desired approximation

$$\underline{v}_n(D^{(\gamma)}) \approx \underline{v}_n(D) + \sum_{\substack{k=1 \\ k \neq n}}^{N} \frac{\langle \Gamma(\gamma)\underline{v}_n(D), \underline{v}_k(D)\rangle}{(\lambda_n - \lambda_k)} \underline{v}_k(D), \tag{6.13}$$

provided that γ is sufficiently small.

The formula in (6.13) is approximate in the sense that we do not have estimates for the error and, instead, we have *assumed* the remainder term is uniformly small in the underlying asymptotic expansion. However, we can verify the accuracy of this formula through simulations, presented in Figure 6.4, where we compare the approximate eigenvector from (6.13) and the true eigenvector for many randomly perturbed cochlea-inspired rainbow sensors. We see that the errors are small when the size of the perturbations γ is small.

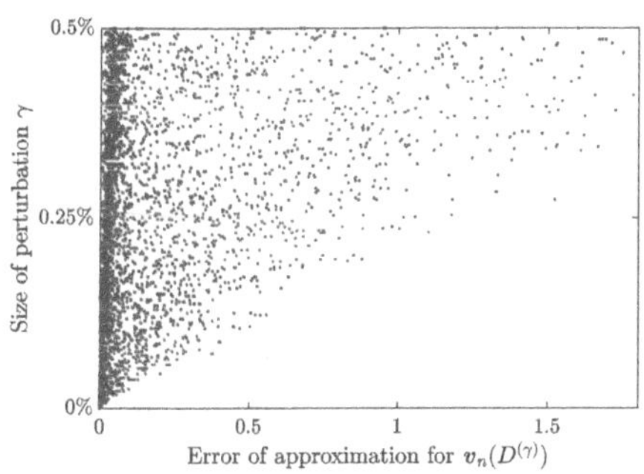

Figure 6.4: The error of the approximation for $\underline{v}_n(D^{(\gamma)})$ derived in (6.13) is small for small perturbations γ, which are expressed as percentages of the unperturbed values. We repeatedly simulate randomly perturbed cochlea-inspired rainbow sensors and compare the exact value with the approximate value from (6.13).

6.3 Removing resonators from the device

We will now consider a particularly drastic class of perturbations of the rainbow sensors: the effect of removing a resonator from the array. This is shown in Figure 6.5. This is inspired by observations of the biological cochlea where in many places the receptor cells are so badly damaged that the stereocilia have been completely destroyed, as depicted in Figure 6.1.

We introduce some notation to describe a system of resonators with one or more resonators removed. Given a resonator array D we write $D^{(i)}$ to denote the same array with the ith resonator removed. The resonators are labeled according to increasing

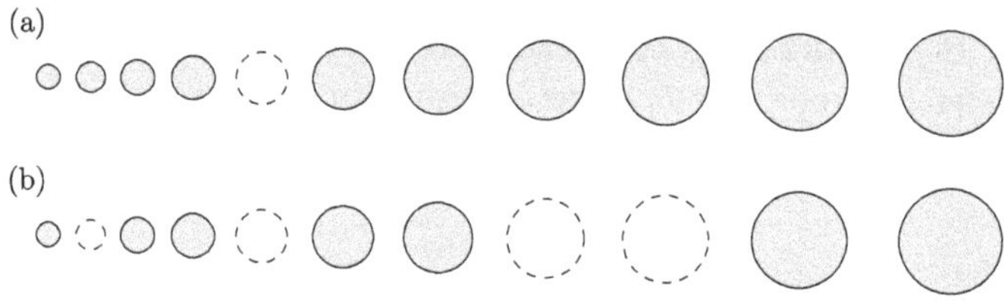

Figure 6.5: We study the effects of removing resonators from a cochlea-inspired rainbow sensor. (a) The rainbow sensor with a single resonator removed, denoted by $D^{(5)}$. (b) The rainbow sensor with multiple resonators removed, denoted by $D^{(2,5,8,9)}$. The original rainbow sensor, $D = D_1 \cup \cdots \cup D_{11}$, is shown in dashes.

volume (so, from left to right in the graded cochlea-inspired rainbow sensors depicted here). For the removal of multiple resonators, we add additional subscripts. For example, in Figure 6.5(a) we show $D^{(5)} = D_1 \cup \cdots \cup D_4 \cup D_6 \cup \cdots \cup D_{11}$ and in Figure 6.5(b) we show $D^{(2,5,8,9)}$, which has the 2nd, 5th, 8th, and 9th resonators removed.

The crucial result that underpins the analysis in this section is Cauchy's interlacing theorem, which describes the relation between a Hermitian matrix's eigenvalues and the eigenvalues of its principal submatrices. A principle submatrix is a matrix obtained by removing rows and columns (with the same indices) from a matrix.

Theorem 6.3.1 (Cauchy's interlacing theorem). *Let A be an $N \times N$ Hermitian matrix with eigenvalues $\lambda_1 \le \lambda_2 \le \cdots \le \lambda_N$. Suppose that B is an $(N-1) \times (N-1)$ principal submatrix of A with eigenvalues $\mu_1 \le \mu_2 \le \cdots \le \mu_{N-1}$. Then, the eigenvalues are ordered such that $\lambda_1 \le \mu_1 \le \lambda_2 \le \mu_2 \le \cdots \le \lambda_{N-1} \le \mu_{N-1} \le \lambda_N$.*

Proof. Various proof strategies exist, see [85] or [91], for example. □

Thanks to Cauchy's interlacing theorem, we can quickly obtain a result for the eigenvalues of the generalized capacitance matrix. In order to state a result for the resonant frequencies of a resonator array, we will first introduce some asymptotic notation.

Definition 6.3.1. For nonnegative real-valued functions f and g, we will write that $f(\delta) \gtrsim g(\delta)$ as $\delta \to 0$ if

$$\lim_{\delta \to 0} \frac{f(\delta)}{\max\{f(\delta), g(\delta)\}} = 1, \quad \text{as } \delta \to 0,$$

where we define the ratio to be 1 in the event that $0 = f = g$.

Lemma 6.3.1. *Let D be a resonator array and $D^{(i)}$ be the same array with the ith resonator removed. Then, if δ is sufficiently small, the resonant frequencies of the two structures interlace in the sense that*

$$\Re(\omega_j(D)) \lesssim \Re(\omega_j(D^{(i)})) \lesssim \Re(\omega_{j+1}(D)) \quad \textit{for all } j = 1, \ldots, N-1.$$

Proof. Since $C(D)$ is symmetric and real-valued, we can use Cauchy's interlacing theorem (Theorem 6.3.1) to see that

$$\lambda_j(D) \leq \lambda_j(D^{(i)}) \leq \lambda_{j+1}(D) \quad \text{for all } j = 1, \ldots, N-1.$$

Then, the result follows from the asymptotic formula in Theorem 4.2.1. □

The subwavelength resonant frequencies of resonator arrays with an increasing number of removed resonators are shown in Figure 6.6. We see that the frequencies interlace those of the previous structure and remain distributed across the audible range. In general, we observe that removing resonators at different parts of the array affects different parts of the spectrum more strongly. If the larger resonators are removed, then the lower frequencies in the spectrum experience the strongest perturbations, while removing the smallest resonators affects the highest frequencies more significantly. This matches the intuition gained from the resonant frequencies of the uncoupled resonators and, crucially, all takes place within the bounds posed by the interlacing property from Lemma 6.3.1.

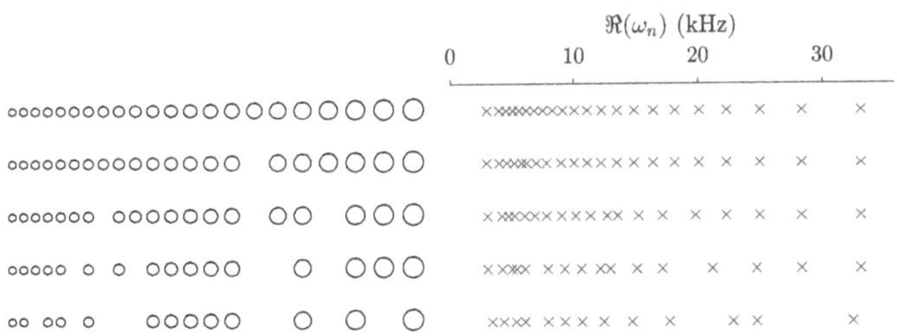

Figure 6.6: The subwavelength resonant frequencies of a cochlea-inspired rainbow sensor with resonators removed. Each subsequent array has additional resonators removed and its set of resonant frequencies interlaces the previous, at leading order, as predicted by Lemma 6.3.1.

In general, Lemma 6.3.1 is useful for understanding the effect of removing a resonator but does not give stability, in the sense of the perturbation being small. However, a cochlea-inspired rainbow sensor with a large number of resonators can be designed such that the resonant frequencies are bounded, even as their number becomes very large. In this case, many of the gaps between the real parts will be small and, subsequently, so will the perturbations caused by removing a resonator. There are a variety of ways to formulate this precisely, one version is given in the following theorem.

Theorem 6.3.2. *Suppose that a resonator array D is dilute with parameter $0 < \epsilon \ll 1$ in the sense that*

$$D = \bigcup_{j=1}^{N} (B + \epsilon^{-1} z_j),$$

where B is a fixed bounded domain and $\epsilon^{-1}z_j$ represents the position of each resonator. Then, there exists a constant $c \in \mathbb{R}$, which does not depend on N or ϵ, such that if $\epsilon = \frac{c}{N}$, then all the eigenvalues $\{\lambda_j\}$ of C^ϵ are such that

$$0 < \lambda_j < \frac{2|\mathrm{Cap}_B|}{|B|}. \tag{6.14}$$

Proof. In this case, the dilute generalized capacitance matrix is given by

$$C_{ij}^\epsilon = \begin{cases} \frac{\mathrm{Cap}_B}{|B|}, & i = j, \\ -\frac{\epsilon \mathrm{Cap}_B^2}{4\pi|B||z_i - z_j|}, & i \neq j. \end{cases} \tag{6.15}$$

By the Gershgorin circle theorem, we know that the eigenvalues $\{\lambda_j : j = 1, \dots, N\}$ must be such that

$$\left|\lambda_j - \frac{\mathrm{Cap}_B}{|B|}\right| \leq \frac{\epsilon \mathrm{Cap}_B^2}{4\pi|B|} \sum_{i \neq j} \frac{1}{|z_i - z_j|}, \quad j = 1, \dots, N. \tag{6.16}$$

Now, we have that

$$\frac{\epsilon \mathrm{Cap}_B}{4\pi} \sum_{i \neq j} \frac{1}{|z_i - z_j|} \leq \epsilon(N-1)\frac{\mathrm{Cap}_B}{4\pi} \sup_{i \neq j} |z_i - z_j|^{-1},$$

which we can choose to be less than 1 by selecting $c = \epsilon N$ appropriately. In which case, we have that the eigenvalues $\{\lambda_j : j = 1, \dots, N\}$ satisfy

$$\left|\lambda_j - \frac{\mathrm{Cap}_B}{|B|}\right| \leq \frac{\mathrm{Cap}_B}{|B|}, \quad j = 1, \dots, N. \qquad \square$$

It is important to note that Theorem 6.3.2 merely shows that the real parts of the resonant frequencies will be bounded, as the number of resonators becomes large. It does not guarantee that they are evenly spaced or that the gaps between any particular adjacent resonant frequencies are small. For example, see Figure 6.7, where the sub-

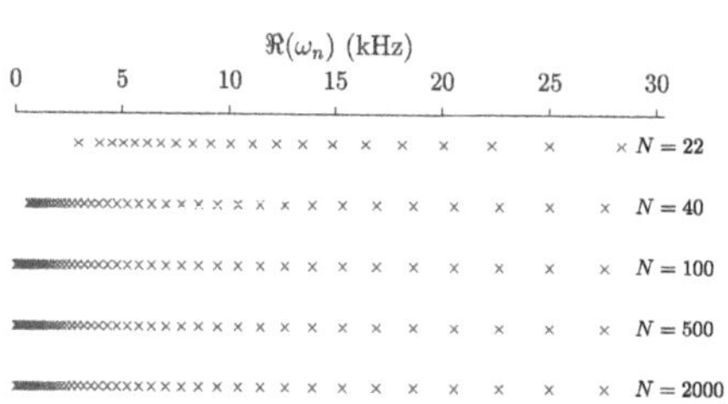

Figure 6.7: Large cochlea-inspired rainbow sensors can be designed such that the subwavelength resonant frequencies are bounded. Here, we simulate successively larger arrays using the multipole expansion to solve the full differential problem, with the configuration of the resonators being chosen according to the dilute regime defined in Theorem 6.3.2.

wavelength resonant frequencies for increasingly large arrays, dimensioned according to Theorem 6.3.2, are shown. We see that the frequencies become very dense in part of the range but remain sparser at higher frequencies.

6.4 Implications for signal processing

Recall from Chapter 5 the link between the cochlea-inspired metamaterials we are considering and bio-inspired signal processing algorithms. It is valuable to consider the implications of the robustness properties studied in the present chapter on these algorithms.

The signal transform that was derived is: given a signal s, compute the N time-varying outputs $a_n[s]$, defined by

$$a_n[s](t) = (s * h[\omega_n])(t), \quad n = 1, \dots, N, \tag{6.17}$$

for kernels defined as

$$h[\omega_n](t) = \begin{cases} 0, & t < 0, \\ c_n e^{\Im(\omega_n)t} \sin(\Re(\omega_n)t), & t \geq 0, \end{cases} \quad n = 1, \dots, N, \tag{6.18}$$

for some real-valued constants c_n. Recall that $\Im(\omega_n) < 0$ due to energy loss, so $h[\omega_n](t)$ decays as $t \to \infty$. We wish to show that the signal transform $s \mapsto a_n[s] := s * h[\omega_n]$ is robust with respect to errors and imperfections in the design of the underlying cochlea-inspired rainbow sensor. Our main result in this regard is the following theorem.

Theorem 6.4.1. *Given two complex numbers ω^{old} and ω^{new} with negative imaginary parts, it holds that*

$$\|s * h[\omega^{\text{old}}] - s * h[\omega^{\text{new}}]\|_{L^\infty(\mathbb{R})} \leq \|h[\omega^{\text{old}}] - h[\omega^{\text{new}}]\|_{L^\infty(\mathbb{R})} \|s\|_{L^1(\mathbb{R})},$$

for all $s \in L^1(\mathbb{R})$.

Proof. This is a standard argument for bounding convolutions:

$$\begin{aligned} \|s * h[\omega^{\text{old}}] - s * h[\omega^{\text{new}}]\|_{L^\infty(\mathbb{R})} &\leq \sup_{x\in\mathbb{R}} \int_{\mathbb{R}} |s(x-y)| |h[\omega^{\text{old}}](y) - h[\omega^{\text{new}}](y)| \, dy \\ &\leq \|h_n^{\text{old}} - h_n^{\text{new}}\|_{L^\infty(\mathbb{R})} \sup_{x\in\mathbb{R}} \int_{\mathbb{R}} |s(x-y)| \, dy \\ &= \|h_n^{\text{old}} - h_n^{\text{new}}\|_{L^\infty(\mathbb{R})} \|s\|_{L^1(\mathbb{R})}. \end{aligned}$$

□

Notice that if s is compactly supported, then we can reframe Theorem 6.4.1 in terms of $\|\cdot\|_{L^p(\mathbb{R})}$ for any $1 \leq p \leq \infty$, using Hölder's inequality.

Corollary 6.4.1. *Let $c > 0$ and suppose we have two complex numbers ω^{old} and ω^{new} whose imaginary parts satisfy $\mathfrak{I}(\omega^{\text{old}}), \mathfrak{I}(\omega^{\text{old}}) \leq -c$. Then, it holds that*

$$\|s * h[\omega^{\text{old}}] - s * h[\omega^{\text{new}}]\|_{L^\infty(\mathbb{R})} \leq \frac{\sqrt{2}}{ce}|\omega^{\text{old}} - \omega^{\text{new}}|\|s\|_{L^1(\mathbb{R})},$$

for all $s \in L^1(\mathbb{R})$.

Proof. We begin with the observation that

$$\begin{aligned}
&|h[\omega^{\text{old}}](t) - h[\omega^{\text{new}}](t)| \\
&= \Big|\big(e^{\mathfrak{I}(\omega^{\text{old}})t} - e^{\mathfrak{I}(\omega^{\text{new}})t}\big)\sin(\mathfrak{R}(\omega^{\text{old}})t) \\
&\quad + e^{\mathfrak{I}(\omega^{\text{new}})t}\big(\sin(\mathfrak{R}(\omega^{\text{old}})t) - \sin(\mathfrak{R}(\omega^{\text{new}})t)\big)\Big|,
\end{aligned}$$

for $t > 0$. Then, we have that

$$\begin{aligned}
\big|\big(e^{\mathfrak{I}(\omega^{\text{old}})t} - e^{\mathfrak{I}(\omega^{\text{new}})t}\big)\sin(\mathfrak{R}(\omega^{\text{old}})t)\big| &\leq \big|e^{\mathfrak{I}(\omega^{\text{old}})t} - e^{\mathfrak{I}(\omega^{\text{new}})t}\big| \\
&\leq \frac{1}{ce}|\mathfrak{I}(\omega^{\text{old}}) - \mathfrak{I}(\omega^{\text{new}})|,
\end{aligned}$$

for $t > 0$, where we have used the fact that $\sup_{t>0}\sup_{\omega<-c}|te^{\omega t}| = \frac{1}{ce}$. Similarly, we have that

$$\big|e^{\mathfrak{I}(\omega^{\text{new}})t}\big(\sin(\mathfrak{R}(\omega^{\text{old}})t) - \sin(\mathfrak{R}(\omega^{\text{new}})t)\big)\big| \leq \frac{1}{ce}|\mathfrak{R}(\omega^{\text{old}}) - \mathfrak{R}(\omega^{\text{new}})|,$$

for $t > 0$, where we have used the fact that $\sup_{t>0}\sup_{\omega<-c}|te^{\omega t}\cos(at)| \leq \frac{1}{ce}$ for any $a \in \mathbb{R}$. Putting this together, we have that

$$\begin{aligned}
&\|s * h[\omega^{\text{old}}] - s * h[\omega^{\text{new}}]\|_{L^\infty(\mathbb{R})} \\
&\leq \frac{1}{ce}\big(|\mathfrak{I}(\omega^{\text{old}}) - \mathfrak{I}(\omega^{\text{new}})| + |\mathfrak{R}(\omega^{\text{old}}) - \mathfrak{R}(\omega^{\text{new}})|\big)\|s\|_{L^1(\mathbb{R})},
\end{aligned}$$

from which we arrive at the result, using the inequality $|a| + |b| \leq \sqrt{2(a^2 + b^2)}$. □

While Theorem 6.4.1 is the standard stability result for convolutional signal processing algorithms, Corollary 6.4.1 is most revealing here. It shows that the outputs of the induced biomimetic signal transform (defined by (6.17) here) are stable with respect to changes in the resonant frequencies of the physical device. From Sections 6.2 and 6.3, we know that the resonant frequencies of the cochlea-inspired rainbow sensor are robust with respect to a variety of errors and imperfections (particularly in large dilute resonator arrays), meaning that the biomimetic signal transform inherits this robustness.

To test the robustness for small arrays with removed resonators, Figure 6.8 shows the frequency support of the filter array used in the biomimetic signal transform in the

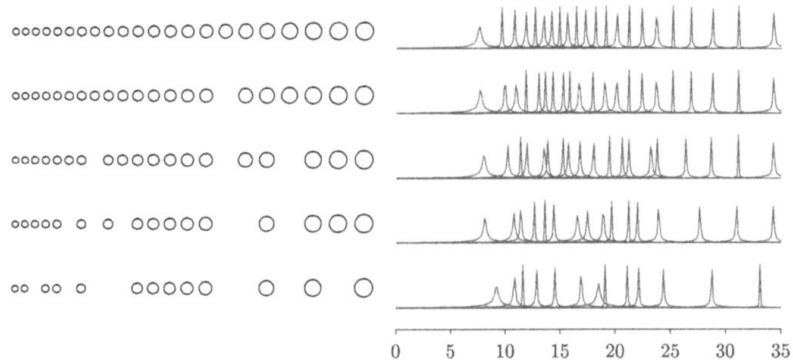

Figure 6.8: The frequency supports of the filter kernels $h[\omega_n]$ induced by a cochlea-inspired rainbow sensor. Each subsequent array has additional resonators removed and for each array we plot the Fourier transform of $h[\omega_n]$, $n = 1, \dots, N$, normalized in $L^2(\mathbb{R})$.

case of successively removed resonators (the same sequence of structures was simulated in Figure 6.6). In this small array (of 22 resonators, initially) we see that gaps emerge when multiple resonators are removed, corresponding to hearing loss at frequencies within these gaps. We observe that even when a significant proportion of resonators are removed from the array, the support of the filter array still covers a large proportion of the range of audible frequencies.

6.5 Discussion

The formulas derived in this chapter show that a cochlea-inspired graded metamaterial is robust with respect to small perturbations in the position and size of the constituent resonators. The effect of removing resonators was also described; it was shown that the change in the subwavelength resonant frequencies can always be bounded via an interlacing theorem and that the resulting change in the spectrum can be small in the case of sufficiently large, dilute arrays. The implication of this analysis for related biomimetic signal transforms were also studied, and it was shown that stability properties are inherited from the underlying resonant frequencies.

The analysis in this work suggests possible mechanisms through which sufficiently large structures could be robust to (surprisingly) large perturbations. However, the extent to which this truly replicates the remarkable robustness of the cochlea is unclear. While the mechanisms which underpin the function of cochlea-inspired rainbow sensors (which are locally resonant graded metamaterials) and biological cochleae (which have a graded membrane with receptor cells on the surface) are quite different, there is scope for further insight to be traded between the two communities.

As a simple example of a comparison that can be drawn between the two classes of systems, we observed a tendency for gaps in the frequency support of the filter array to appear more commonly at higher frequencies (see Figure 6.8, for example) [64]. This is (qualitatively) consistent with the observation that human hearing loss initially occurs at high frequencies in most people [169]. It is not yet clear if either of these ideas can be transposed between the two realms of biological hearing models and cochlea-inspired devices.

An example of a particularly relevant breakthrough in the field of metamaterials has been the development of *topologically protected* systems. This principle exploits the underlying topological properties of periodic structures to develop wave systems whose properties are robust with respect to imperfections and defects [104]. This robustness is a consequence of the fact that the wave guiding properties of the material are due to the topological classifications of the underlying periodic materials. Hence, so long as the imperfections and defects do not have the effect of changing the topological classification of the material, the desired wave guiding behavior will remain [75].

The canonical example of topologically nontrivial wave dynamics in wave systems that are periodic in one dimension is the Su–Schrieffer–Heeger (SSH) model and variants thereof. Originally developed for modeling electrons in polyacetylene (a polymer), the model features alternating weak and strong coupling strengths between atoms [153]. In this case, the appropriate topological index is the *Zak phase* and it has been shown that creating interfaces between materials with different Zak phases leads to wave localization that is "topologically protected" in the sense that it is robust with respect to imperfections and defects [172]. Many variants have been considered in wave physics, e. g., [104, 142, 171], including systems of high-contrast subwavelength resonators [12]. In particular, it was shown in [8] how to fine tune a topologically protected array of subwavelength resonators in order to optimize the robustness.

Notions of topological protection have also been extended to graded systems and rainbow effects. For example, graded SSH systems have been developed, taking advantage of topologically nontrivial band gaps to perform the spatial frequency separation [49]. These ideas have also influenced the design of wave-based signal processing devices, leading to substantial improvements in the robustness of devices [173]. On top of this, the insight generated by notions of topological protection have lead to substantial breakthroughs in the world of graded metamaterials. For example, it has been shown that by breaking symmetries in the system, in a topologically nontrivial way, zero group velocity modes can be induced inside the Brillouin zone [51]. This is in contrast with the systems considered here for which the edges of the band gaps always occurred at the edges of the Brillouin zone (as shown in Figures 2.2 and 2.3, for example). It was shown in [51] that having zero group velocity modes within the Brillouin zone leads to longer interaction times and increases the energy harvested from a wave (which was the application they had in mind). Whether this increase in interaction time would have implications for the signal processing action performed by the graded metamaterial is not yet clear, but is an exciting question for future investigation.

7 Active metamaterials with nonlinear amplification

In this final chapter, we will consider adding active elements to the cochlea-inspired metamaterial. This is intended to replicate the nonlinear amplification mechanism that is used by the biological cochlea.

Humans are also able to comfortably hear sounds with amplitudes that range over an impressive six orders of magnitude. This is thanks to the cochlea's nonlinear amplification response, known as the *cochlear amplifier*. This is such that a compressive nonlinearity occurs, whereby quieter sounds are amplified much more greatly than louder ones. Understanding the details of the mechanism responsible for the cochlear amplifier represents the most significant open problem in the study of hearing. Recently, the most prevalent models have been based on a critically-poised Hopf resonator, the details of which are discussed below.

There are a wide range of active elements that have been implemented in metamaterials [170, 41]. Several notable examples of simple strategies rely on mechanically deforming a substrate or matrix to alter the proximity of crystalline elements [138, 164, 73, 135]. Alternative methods rely on using materials that respond to an external stimulus, such as thermo-mechanical excitation [154], electro-mechanical excitation [33], and photo-responsive excitation [50] control, among others. In the specific example of graded acoustic metamaterials, this has been done using coupled speakers and microphones in a coupled resonator array [146].

7.1 Cochlear amplification

At its simplest, the cochlea is a long tube that is filled with fluid and divided in two by the cochlear partition [59]. This partition is set in motion by acoustic waves in the cochlear fluid and contains receptor cells which respond to this motion [88, 143, 120]. These cells, known as *hair cells*, can be split into two categories, referred to as *inner* and *outer* due to their positioning. These receptor cells were visible in Figure 6.1. Inner hair calls are responsible for converting motion of the partition into electrical signals, while outer hair cells contain a motor protein and are the facilitator of the cochlea's nonlinear amplification mechanism.

While there is still much to be learnt about the cochlear amplifier, researchers are guided by a handful of valuable observations. It is understood, firstly, that this mechanism acts as a negative-damping mechanism [84, 130, 96]. Secondly, it has been observed that nonlinear amplification occurs with a one-third power law, suggesting the presence of a cubic nonlinearity [89].

Hopf resonators have become popular objects to study in the field of nonlinear cochlear mechanics thanks to their remarkable ability to account for the key properties that typify cochlear behavior [74, 89, 90, 46, 96, 72, 103, 110, 117]. The normal form

https://doi.org/10.1515/9783110784961-007

of a single Hopf resonator $z = z(t) : \mathbb{R} \to \mathbb{C}$ in the complex plane is given by the forced differential equation

$$\frac{\mathrm{d}z}{\mathrm{d}t} = (\mu + \mathrm{i}\omega_0)z - \beta|z|^2 z + F, \tag{7.1}$$

where $F = F(t)$ is the forcing term and μ, ω_0 and $\beta > 0$ are real parameters. This system is a resonator in the sense that the absolute value of the response z is greatest when the forcing F occurs with frequency ω_0. In cochlear models, z is some variable which characterizes the system's state.

The parameter μ is the bifurcation parameter. For $\mu < 0$, the unforced system ($F = 0$) has a stable equilibrium at $z = 0$ whereas when $\mu > 0$ this equilibrium is unstable and there exists a stable limit cycle given by $z(t) = \sqrt{\mu/\beta}e^{\mathrm{i}\omega_0 t}$. This birth of a limit cycle is typical of a (supercritical) Hopf bifurcation, which is formally characterized by a conjugate-pair of linearized eigenvalues crossing the imaginary axis. Writing the unforced system (7.1) in terms of its real and imaginary parts and linearizing about the fixed point at zero gives a system whose Jacobian matrix has eigenvalues $\lambda = \mu \pm \mathrm{i}\omega_0$. These eigenvalues clearly cross the imaginary axis when μ passes zero. For further details, see, e. g., [152, 93, 163].

The greatly enhanced response for frequencies close to ω_0 is able to account for the cochlea's frequency selectivity. The cubic nonlinearity in (7.1) is able to reproduce the one-third power law of the cochlea: when μ is small, and the system is close to bifurcation, we have that $|z| \approx |F|^{1/3}$ for frequencies close to resonance. One of the earliest pieces of evidence supporting the active nature of the cochlea was the observation that the ear emits sounds (known at otoacoustic emissions) as part of its response [102, 177]. The existence of stable limit cycles, for certain parameter values, predicts this behavior [46, 72, 100].

A further symptom of the nonlinearity that exists in the cochlea is the behavior that is observed under the influence of a signal composed of two distinct tones. Firstly, when the ear is excited by such a stimulus two-tone suppression occurs. That is, the frequency spectrum of the response contains the expected two amplitude peaks, however, these are smaller than each would be in the absence of the other tone [145]. Further, in this situation the ear also detects additional tones, variously known as combination tones, distortion products or Tartini's tones [86, 144, 97]. Close to bifurcation, the nonlinearity in (7.1) gives products that can account for these phenomena [97, 72].

In this chapter, we will introduce a Hopf-type nonlinearity directly to the wave-propagation problem by supposing that the resonators are equipped with an appropriate forcing mechanism. For simplicity, we will use the two-dimensional model considered in Chapter 3, construct an eigenmode decomposition and add a nonlinear forcing term. We will explore the Hopf-type behavior of this system and show that the crucial cochlea-like properties of (7.1) are retained by the coupled subwavelength structure.

7.2 Nonlinear amplification

We will use a modal decomposition to analyze the wave propagation within the structure [163, 79]. That is, we wish to express the acoustic pressure $p = p(x,t)$ at position x and time t in the form

$$p(x,t) = \mathbb{R}\left(\sum_n \alpha_n(t)u_n(x)\right). \tag{7.2}$$

Separating variables in the unforced, linear wave equation yields the spatial Helmholtz problem given by (3.3). In Chapter 3 we saw how boundary integral formulations allowed us to perform both asymptotic and numerical analyses, in order to characterize the resonant modes of the system.

We now wish to introduce appropriate nonlinear amplification to the model. As discussed above, in Section 7.1, the canonical form of a Hopf resonator is able to account for the important properties of the cochlear amplifier. This suggests adding amplification based on a nonlinearity of the form

$$\mathcal{N}[\varphi] := \mu\varphi - \beta|\varphi|^2\varphi. \tag{7.3}$$

These two terms, respectively, account for the negative damping and cubic non-linearity that we said our amplification should include.

An important consideration, when choosing to introduce amplification, is the stability of the system. For example, Rupin et al. [146] used a formulation whereby amplification closely resembling $\mathcal{N}[p]$ was added. In order for this formulation to be stable, it was necessary to design a set-up that switched off the amplification if the pressure exceeded a threshold value. Conversely, there exist a number of formulations which are stable without this thresholding. Examples include variants of $\mathcal{N}[\partial_t p]$ used in the artificial cochlear devices of Joyce and Tarazaga [96, 94, 95] and the $\mu\partial_t p - \beta|p|^2\partial_t p$ term considered by Duke and Jülicher [72].

In this work we will study the system produced by introducing amplification of the form $\mathcal{N}[\partial_t p]$ to the resonators. As we shall see below, this system is stable without the need to impose a pressure threshold. Furthermore, there is evidence which suggests that hair cell stimulation (by stereocilia displacement) is dependent on membrane velocity [114], suggesting that any amplification should be a function of $\partial_t p$.

Thus, we are interested in solving the forced, amplified wave-propagation problem

$$\left(\frac{\partial^2}{\partial t^2} - c(x)^2\Delta\right)p(x,t) = f(t)\mathcal{X}_Q(x) + \mathcal{N}\left[\frac{\partial}{\partial t}p(x,t)\right], \tag{7.4}$$

where $c(x) = v\mathcal{X}_D(x) + v_0\mathcal{X}_{\mathbb{R}^2\setminus\overline{D}}(x)$ and $Q \subset \mathbb{R}^2$ is a compact set on which the forcing is applied.

Since the N subwavelength modes are expected to dominate the response to an audible frequency, we truncate the expansion (7.2) and seek a solution of the form

$$p(x,t) = \mathbb{R}\left(\sum_{n=1}^{N} a_n(t)u_n(x)\right), \tag{7.5}$$

for some time-dependent coefficients $a_1(t), \dots, a_N(t)$. For some of the analysis in this section, we will assume that the amplitude of the pressure variation is small, such that the cubic term does not dominate. In which case, we may, as a reasonable approximation, use the linear eigenmodes u_n from Chapter 3 in the expansion (7.5) [163]. We will also project the forcing term onto the space spanned by the subwavelength modes. Define the matrix $\Gamma \in \mathbb{C}^{N\times N}$ as $\Gamma_{ij} := (u_i, u_j)_{2,Q}$, where $(\cdot,\cdot)_{2,Q}$ is the standard inner product on $L^2(Q)$. An important property is that thanks to the linear independence of the eigenmodes, Γ is invertible, as discussed in Lemma 3.6.1. We can then decompose the forcing as

$$f(t)\mathcal{X}_Q(x) \simeq f(t)\sum_{n=1}^{N} \mathcal{F}_n u_n(x), \tag{7.6}$$

where the coefficients $\mathcal{F}_n$ are given by

$$\begin{pmatrix} \mathcal{F}_1 \\ \vdots \\ \mathcal{F}_N \end{pmatrix} = \Gamma^{-1} \begin{pmatrix} (\mathcal{X}_Q, u_1)_{2,Q} \\ \vdots \\ (\mathcal{X}_Q, u_N)_{2,Q} \end{pmatrix}. \tag{7.7}$$

We now substitute the ansatz (7.5) into the forced wave equation (7.4). In light of the fact that $c(x)^2\Delta u_n(x) = -\omega_n^2 u_n(x)$ (and that each u_n satisfies the appropriate transmission properties across ∂D), we reach the problem

$$\sum_{n=1}^{N}(a_n''(t) + \omega_n^2 a_n(t))u_n(x) = f(t)\sum_{n=1}^{N}\mathcal{F}_n u_n(x) + \mathcal{N}\left[\sum_{n=1}^{N} a_n'(t)u_n(x)\right]\mathcal{X}_D(x). \tag{7.8}$$

Our approach to studying (7.8) will be to take the $L^2(D)$ product with u_m, for $m = 1, \dots, N$, to reach a coupled system of N ordinary differential equations. Define the matrix $\gamma \in \mathbb{C}^{N\times N}$ as $\gamma_{ij} := (u_i, u_j)_{2,D}$, which is known to be invertible (Lemma 3.6.1). If we write $a_1, \dots, a_N$ in the column vector $\boldsymbol{a}$ then the modal system is described by

$$\boldsymbol{a}'' + \boldsymbol{\Lambda}\boldsymbol{a} - \mu\boldsymbol{a}' + \beta\mathbf{N}(\boldsymbol{a}') = f(t)\mathbf{F}, \tag{7.9}$$

where $\boldsymbol{\Lambda} \in \mathbb{C}^{N\times N}$ is diagonal with entries ω_n^2, $\mathbf{F} \in \mathbb{C}^N$ is a vector of the forcing constants $\mathcal{F}_1, \dots, \mathcal{F}_N$ and $\mathbf{N} : \mathbb{C}^N \to \mathbb{C}^N$ is the nonlinear function defined as

$$\mathbf{N}(\mathbf{z}) := \gamma^{-1}\left[\int_D \left|\sum_{n=1}^{N} z_n u_n(x)\right|^2 \left(\sum_{n=1}^{N} z_n u_n(x)\right)\overline{u_j(x)}\,\mathrm{d}x\right]_{j=1,\dots,N}. \tag{7.10}$$

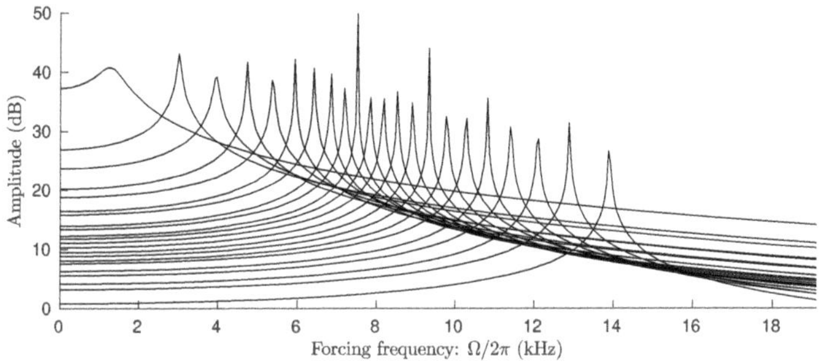

Figure 7.1: The response of each amplitude $|X_1|, \ldots, |X_N|$ (with peaks from left to right) as a function of the incident frequency Ω. Each mode experiences a peak of excitation in the vicinity of its resonant frequency. The incident sound is at 100 dB SPL.

In much of the following analysis, we will be interested in the case when $f(t)$ is harmonic with frequency Ω. In this case, we can approximate the solution to (7.9) using a harmonic balance approach [163, 151, 79, 128]. That is, if $f(t) = Fe^{-i\Omega t}$, for $F, \Omega \in \mathbb{R}$, then we may approximate the steady-state solutions to (7.9) as $a_k(t) = X_k e^{-i\Omega t + i\psi_k}$ for amplitudes $X_k \in \mathbb{R}$ and phase delays $\psi_k \in \mathbb{R}$. Making this substitution leads to a system of coupled cubic equations that can be solved numerically.

In Figure 7.1 we see that, as is to be expected, that as the forcing frequency is varied each mode is excited much more greatly in the vicinity of the associated resonant frequency (in spite of the coupling within the nonlinearity (7.10)). This motivates an approximate system whereby, if the system is forced at a frequency close to one of the resonant frequencies, we assume that only that mode is excited.

7.3 Single-mode approximation

When the forcing frequency Ω is close to one of the resonant frequencies ω_k, we can approximate the solution by assuming that only the corresponding mode u_k is excited. Under such an assumption (that $a_n = 0$ if $n \neq k$) taking the $L^2(D)$ product of (7.8) with u_k yields a simplified version of (7.9), given by

$$a_k'' + \omega_k^2 a_k - \mu a_k' + \hat{\beta}|a_k'|^2 a_k' = f(t)\mathcal{F}_k, \tag{7.11}$$

where $\hat{\beta} := \beta\|u_k\|_{4,D}^4/\|u_k\|_{2,D}^2$ (and $\|\cdot\|_{p,D}$ is the $L^p(D)$-norm).

At this point, we pause to explore the Hopf-type behavior that is exhibited by our model. In the case that $f = 0$, we see that (7.11) has a periodic solution $a_k(t) = R_k^c e^{-i\Omega_k^c t}$ provided that $\mu \geq \mu_k^c$, where

$$\Omega_k^c := \sqrt{\mathfrak{R}(\omega_k)^2 - \mathfrak{I}(\omega_k)^2},$$
$$\mu_k^c := \frac{-2\mathfrak{R}(\omega_k)\mathfrak{I}(\omega_k)}{\Omega_k^c}, \quad R_k^c = \sqrt{\frac{\mu - \mu_k^c}{\hat{\beta}}\frac{1}{\Omega_k^c}}. \tag{7.12}$$

This birth of a limit cycle is typical of a Hopf bifurcation. A Hopf bifurcation is characterised by a conjugate pair of linearised eigenvalues crossing the imaginary axis [152]. Decomposing α_k into its real and imaginary parts, we can write (7.11) as a four-dimensional system of first-order ordinary differential equations. Linearizing this system around the fixed point at $\alpha_k = 0$ gives the Jacobian matrix

$$J = \begin{bmatrix} 0 & 0 & 1 & 0 \\ 0 & 0 & 0 & 1 \\ -(\Omega_k^c)^2 & \mu_k^c\Omega_k^c & \mu & 0 \\ -\mu_k^c\Omega_k^c & -(\Omega_k^c)^2 & 0 & \mu \end{bmatrix}, \tag{7.13}$$

which has eigenvalues given, using the notation of (7.12), by

$$\lambda = \frac{1}{2}\left(\mu \pm \sqrt{-4\left(\Omega_k^c\right)^2 \pm 4\mathrm{i}\Omega_k^c\mu_k^c + \mu^2}\right). \tag{7.14}$$

When $\mu = \mu_k^c$, the eigenvalues of J are $\lambda = \pm\mathrm{i}\Omega_k^c, \mu_k^c \pm \mathrm{i}\Omega_k^c$. It can also be shown that $\frac{\mathrm{d}}{\mathrm{d}\mu}\mathfrak{R}(\lambda) > 0$ meaning that the pair of eigenvalues cross the imaginary axis (from left to right) as μ passes the critical value.

In order to visualize the local stability of this limit cycle and the fixed point at $\alpha_k = 0$, we allow the radius R_k to be a slowly varying function of t. That is, we use the ansatz $\alpha_k = R_k(t)e^{-\mathrm{i}\Omega_k^c t}$ in (7.11) and disregard any terms containing either R_k'' or products of R_k'. This approach leads to the equation

$$\left(-2\mathrm{i}\Omega_k^c R_k' - \left(\Omega_k^c\right)^2 R_k + \omega_k^2 R_k\right) - \mu\left(R_k' - \mathrm{i}\Omega_k^c R_k\right) + \hat{\beta}\left(\left(\Omega_k^c\right)^2 R_k^2 R_k' - \mathrm{i}\left(\Omega_k^c\right)^3 R_k^3\right) = 0, \tag{7.15}$$

which is linear in R_k'. The phase planes for $\mu > \mu_k^c$ and $\mu < \mu_k^c$ (Figure 7.2) demonstrate that as μ passes the critical value a stable limit cycle (described by (7.12)) is born out of the stable equilibrium at the origin, as is typical of a (supercritical) Hopf bifurcation.

An important consideration, when choosing an appropriate nonlinearity, is the stability of the system. We explore the stability of unforced solutions to (7.11) using a technique known as averaging [163, 141, 128]. The above analysis showed that there is a locally stable limit cycle (when $\mu > \mu_k^c$) but it is valuable to understand what happens if $R_k(0)$ is further away from R_k^c. We begin with the ansatz

$$\alpha_k(t) = R_k(t)e^{-\mathrm{i}\Omega_k^c t + \mathrm{i}\psi_k(t)}, \quad \alpha_k'(t) = -\mathrm{i}\Omega_k^c R_k(t)e^{-\mathrm{i}\Omega_k^c t + \mathrm{i}\psi_k(t)}. \tag{7.16}$$

We assume for this analysis that $\hat{\beta} =: \epsilon > 0$ is small and that $\mu = \mu_k^c + \epsilon$.

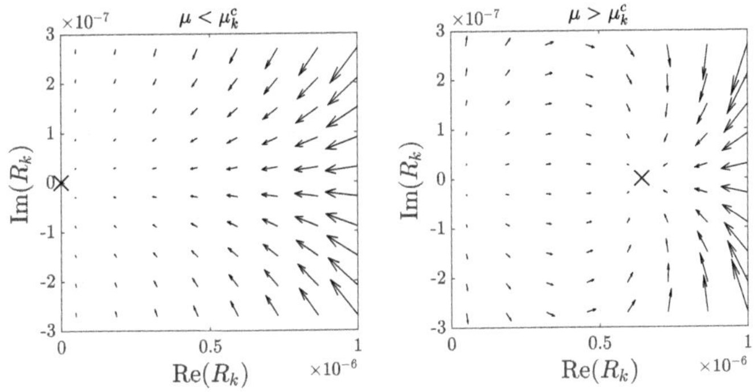

Figure 7.2: The birth of a limit cycle at Hopf bifurcation. For $\mu < \mu_k^c$ the origin is a stable equilibrium. When $\mu > \mu_k^c$ a stable limit cycle with radius R_k^c is born. We depict $\mu = \mu_k^c \pm 100$ and show the stable equilibria with crosses. Here, $k = 11$ and $\hat{\beta} = 10^5$ S Pa^{-2}.

Differentiating the first expression of (7.16) and substituting into the second yields

$$(R_k' + \mathrm{i}\psi_k' R_k)e^{-\mathrm{i}\Omega t + \mathrm{i}\psi_k} = 0. \tag{7.17}$$

Further, substituting (7.16) into (7.11) gives

$$(-\mathrm{i}\Omega_k^c R_k' + \Omega_k^c \psi_k' R_k)e^{-\mathrm{i}\Omega t + \mathrm{i}\psi_k} = \epsilon \mathrm{i}(-\Omega_k^c R_k + (\Omega_k^c)^3 R_k^3)e^{-\mathrm{i}\Omega t + \mathrm{i}\psi_k}. \tag{7.18}$$

We may take the real parts of (7.17) and (7.18) and solve for R_k' and ψ_k' to give

$$R_k' = \epsilon(R_k - (\Omega_k^c)^2 R_k^3)\sin^2(\Omega_k^c t - \psi_k), \tag{7.19}$$

$$\psi_k' = \epsilon(-R_k + (\Omega_k^c)^2 R_k^3)\sin(\Omega_k^c t - \psi_k)\cos(\Omega_k^c t - \psi_k). \tag{7.20}$$

We now make a near-identity transformation in order to express R_k and ψ_k in terms of their average values over the interval $(t - \pi/\Omega_k^c, t + \pi/\Omega_k^c)$, which we denote by $\tilde{R}$ and $\tilde{\psi}$. This transformation has the form

$$R_k = \tilde{R} + \epsilon h_1(\tilde{R}, \tilde{\psi}, t) + O(\epsilon^2), \tag{7.21}$$

$$\psi_k = \tilde{\psi} + \epsilon h_2(\tilde{R}, \tilde{\psi}, t) + O(\epsilon^2), \tag{7.22}$$

where h_1 and h_2 should be chosen in order to simplify the equations for $\tilde{R}$ and $\tilde{\psi}$ as much as possible. This substitution leads to the equations

$$\tilde{R}' = \epsilon\left(-\frac{\partial h_1}{\partial t} + (\tilde{R} - (\Omega_k^c)^2 \tilde{R}^3)\sin^2(\Omega_k^c t - \tilde{\psi})\right) + O(\epsilon^2), \tag{7.23}$$

$$\tilde{\psi}' = \epsilon\left(-\frac{\partial h_2}{\partial t} + (-\tilde{R} + (\Omega_k^c)^2 \tilde{R}^3)\sin(\Omega_k^c t - \tilde{\psi})\cos(\Omega_k^c t - \tilde{\psi})\right) + O(\epsilon^2). \tag{7.24}$$

Ideally, we would like to choose h_1 so that it cancels with the other $O(\epsilon)$ term in (7.23). However, this antiderivative might grow in time meaning the expansion (7.21) will not be valid for large t. Instead, we take h_1 as the antiderivative minus a linear term that grows with the average value [141], that is,

$$h_1(\tilde{R},\tilde{\psi},t) = \int_0^t (\tilde{R} - (\Omega_k^c)^2\tilde{R}^3)\sin^2(\Omega_k^c t - \tilde{\psi})\,\mathrm{d}t - \left[\frac{\Omega_k^c}{2\pi}\int_0^{2\pi/\Omega_k^c} (\tilde{R} - (\Omega_k^c)^2\tilde{R}^3)\sin^2(\Omega_k^c t - \tilde{\psi})\,\mathrm{d}t\right] t. \tag{7.25}$$

After substitution of (7.25) into (7.23), we make an approximation in the spirit of the "averaging" methodology [163, 141, 128]. We will assume that the integral in the second term of (7.25) can be well approximated by taking the value of $\tilde{R}$ and $\tilde{\psi}$ as constant over a cycle of oscillation, leaving a simple trigonometric integral.

We choose h_2 similarly and find that, up to an error of order $O(\epsilon^2)$,

$$\tilde{R}' = \frac{1}{2}\epsilon(\tilde{R} - (\Omega_k^c)^2\tilde{R}^3), \quad \tilde{\psi}' = 0. \tag{7.26}$$

Solving by separation of variables gives that

$$R_k(t) = \frac{1}{\sqrt{R_k(0)^{-2}e^{-\epsilon t} + (\Omega_k^c)^2(1 - e^{-\epsilon t})}} + O(\epsilon). \tag{7.27}$$

Crucially, for any $R_k(0) > 0$ it holds that $R_k(t) \to R_k^c$ as $t \to \infty$, demonstrating that this limit cycle is asymptotically stable.

Consider the case of an incoming signal that consists of a single pure tone at frequency Ω, that is, $f(t) = Fe^{-\mathrm{i}\Omega t}$ for $F, \Omega \in \mathbb{R}$, where Ω is close to ω_k. Using the harmonic balance ansatz $a_k(t) = R_k e^{-\mathrm{i}\Omega t + \mathrm{i}\psi_k}$ and finding the complex modulus of the resulting equation, we arrive at the amplitude–frequency response relation

$$((\Omega_k^c)^2 - \Omega^2)^2 R_k^2 + (-\mu_k^c\Omega_k^c R_k + \mu\Omega R_k - \hat{\beta}\Omega^3 R_k^3)^2 = F^2|\mathcal{F}_k|^2. \tag{7.28}$$

There is a sharply increased response when Ω is close to the resonant frequency associated with the eigenmode, as seen in Figure 7.3. Different magnitudes of force F are shown. When the force is smaller, the response is much greater, thereby allowing the model to capture a very large range of forcing amplitudes with only relatively small variations in acoustic pressure.

We can also observe (by solving for ψ_k) that a phase delay of half a cycle is accumulated as we cross the resonant frequency. The group delay, the time required for information to be delivered, is then given by the derivative $-\mathrm{d}\psi/\mathrm{d}\Omega$. It is predicted that delays of several milliseconds are observed in the vicinity of resonance (Figure 7.3).

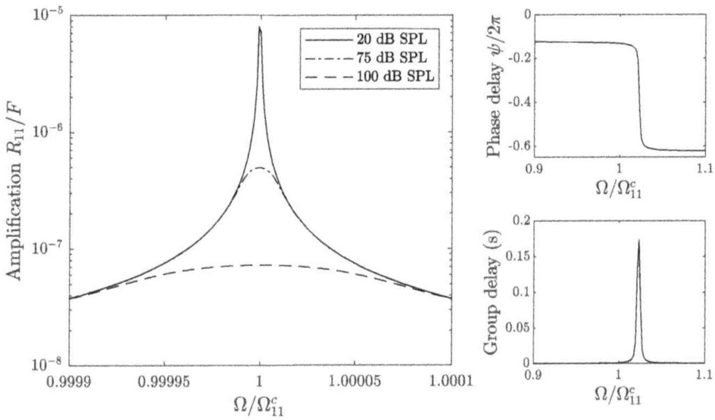

Figure 7.3: The nonlinear response of the single-mode system. Amplification scales nonlinearly with amplitude, and is greater for quieter sounds. Close to resonance, a phase delay of half a cycle is accumulated as well as a sharp increase in group delay. We take $\hat{\beta} = 10^5\,\mathrm{s\,Pa^{-2}}$ and $\mu = \mu^c_{11}$ so that the system is poised at bifurcation. The delay plots are shown for 20 dB SPL.

When studying relations such as (7.28), it becomes apparent that the resonant behavior occurs slightly away from where it is expected (based on the linear system). This is a general property of nonlinear systems and can be understood by examining the harmonic response of the unforced nonlinear system. Solving (7.28) in the case $F = 0$ gives the relationship shown in Figure 7.4, known as a *backbone curve* [163], which described how the natural harmonic response of the nonlinear system is perturbed away from the resonant frequency with increasing amplitude.

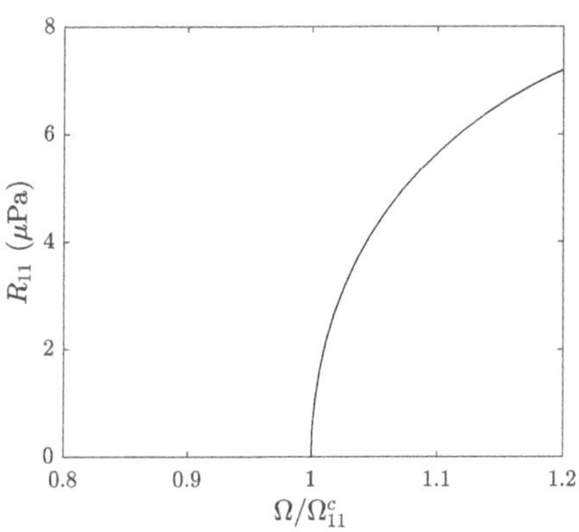

Figure 7.4: The backbone curve of the single-mode equation. Thanks to the non-linearity, the natural response frequency varies as a function of the amplitude. $\hat{\beta} = 10^5\,\mathrm{s\,Pa^{-2}}$ and $\mu = \mu^c_{11}$.

7.4 Fully-coupled system

Bearing in mind the above analysis of the single-mode approximation (7.11), we now return to the fully-coupled system (7.9). Much of the analysis from Section 7.3 can be readily repeated for the matrix system, particularly with the use of numerical schemes for solving nonlinear systems of equations, as was used to produce Figure 7.1. We focus our attention on the elements which tangibly differ from the above discussions.

If we repeat the linear eigenvalue analysis we find that a series of Hopf bifurcations take place, at successive parameter values. When we linearize (7.9) about $\boldsymbol{\alpha} = 0$, since the coupling between modes takes place within the nonlinear part of the system, we reach N uncoupled linear systems each of which has a Jacobian of the form (7.13). This means that each time μ passes one of the critical values μ_n^c, $n = 1, \dots, N$ (as defined in (7.12)), a Hopf bifurcation occurs.

An important feature of the fully-coupled system, which we highlight since it is not the case for the single-mode formulation, is the ability to predict phase delays of more than half a cycle. In uncoupled oscillator systems the phase delay will not exceed half a cycle (cf. Figure 7.3), however, the cochlea is well known to exhibit delays of several cycles [34, 143]. Recalling the decomposition (7.5) and the harmonic balance techniques used above, the phase delay at the point x_0 is given by the complex argument

$$\arg\Big(\sum_n R_n e^{i\psi_n} u_n(x_0)\Big). \tag{7.29}$$

An example of how (7.29) varies as a function of the harmonic forcing frequency is shown in Figure 7.5. The alternating cliffs and plateaus are because the delay increases much more quickly in the region of one of the system's resonant frequencies. This is notably different from the smooth curve corresponding to a biological cochlea and is

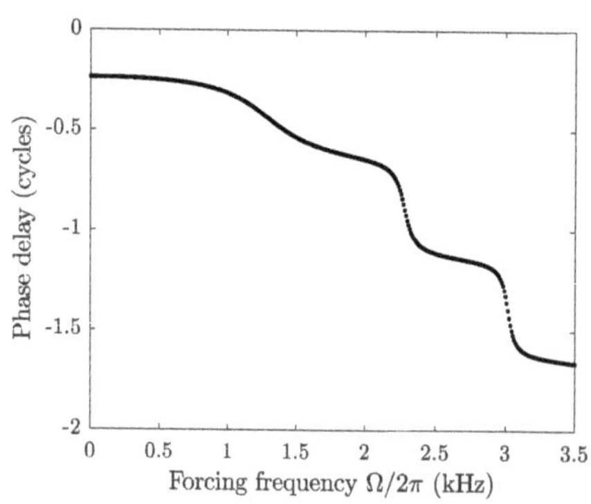

Figure 7.5: The phase delay can, in the fully-coupled system, accumulate to several cycles as the forcing frequency is increased. We study the solution at the center of the 3rd resonator in response to a sound at 100 dB SPL.

a consequence of the discrete nature of a graded metamaterial (which is necessarily composed of discrete meta-atoms).

7.5 Discussion

We have explored, in an abstract theoretical setting, the potential to add nonlinear amplification to the cochlea-inspired graded metamaterials considered in the previous chapters of this book. We introduced a nonlinear forcing term that was designed to replicate the key features of a critically-poised Hopf resonator. This is a popular model in the literature for cochlear amplification and means that our system replicates many of the key features of the cochlear amplifier.

Although implementation of nonlinear amplifications in metamaterials is a challenging experimental problem, it has been done successfully for scaled-up systems, e. g., by [146]. Replicating this at similar dimensions to the cochlea, for example by utilizing the subwavelength resonators considered here, is an important next step for the community.

8 Conclusions and outlook

We have presented a to-scale design for an acoustic metamaterial that is capable of mimicking the properties of the cochlea. Based on a size-graded array of high-contrast resonators, the structure has similar dimensions to the cochlea and has a resonant spectrum that falls broadly within the range of audible frequencies. This design is able to filter different frequencies in space and, with the introduction of a nonlinear amplification term, replicate the fundamental properties of the cochlear amplifier.

The high-contrast model problem considered in this book is motivated by arrays of gas-filled resonators surrounded by an unbounded expanse of fluid. Through this simple setup, it has been demonstrated that with only a few carefully chosen features (a large material contrast, size grading and nonlinear amplification) we are able to design a device capable of replicating the response of the cochlea at a similar scale.

The boundary integral methods presented in this book facilitated both numerical modeling (through multipole expansions) and asymptotic analysis (in terms of the generalized capacitance matrix). While the numerical examples presented here all used either circular or spherical resonators, a major advantage of the integral methods is that they can handle a broad class of (Hölder continuous) boundaries, so could be used to develop new designs with potentially exotic shapes of resonators. It would also be possible, through an appropriate modification of the Green's function, to relax the assumption of open boundary conditions and consider setups with fixed boundaries, for example.

It is valuable to consider some next steps for the development of cochlea-inspired graded metamaterials. Clearly, nonlinearity is a crucial feature of the cochlear response. The human auditory system would not be able to capture such wide ranges of sound levels so efficiently and with the same frequency resolution with a linear response. Many of the artificial devices and metamaterials that have been developed are based on linear responses, for simplicity. We showed, in Chapter 7, the theoretical potential of using active metamaterials to better mimic the cochlear response. However, experimental realizations of these ideas have so far been limited, due to the substantial challenges posed by their implementation [61, 95, 96, 146]. With recent breakthroughs in the development of active and reconfigurable metamaterials [41, 170], there is scope for developing nonlinear systems that would better replicate cochlear function and benefit from enhanced dynamic scaling and frequency resolution.

While most cochlear models are one dimensional, and the systems we considered in this book all featured graded resonators arranged in a straight line, multidimensionality needs to be considered in cochlear models. While the cochlea separates frequencies along one axis, it is a three-dimensional structure. There is evidence that the multidimensional nature of the cochlea is important, as highlighted by [58] which shows that a spiral shape alters the frequency separation of a graded material. It is important that future models account for this multidimensionality and that metamaterials and devices leverage properties such as the spiral to enhance functionality.

https://doi.org/10.1515/9783110784961-008

The interdisciplinary exchanges discussed in this book (and depicted in Figure 1.2) present some exciting opportunities that have not yet been exploited to their full potential. For example, many opportunities exist for developing cochlear understanding by studying mimetic devices. This strategy has been exploited in the past by researchers including Helmholtz [32] and Riemann [35]. Since the metamaterial devices discussed in this book are reaching sufficient maturity that opportunities are emerging, it may be time to resume this line of enquiry. If a researcher builds a cochlea-like device in the lab, it may teach them about how the cochlea functions and serve as a convenient platform for experimentation. Even a simple model, such as one with two cavities and a graded elastic membrane suspended between them, could reveal new insight on issues such as the role of the spiral and the existence of backwards traveling waves, for example.

There is also significant potential for making substantial contributions to the development of signal processing algorithms (and machine hearing approaches in general) by exploiting the recently breakthroughs in graded metamaterial design. The possibility to use the principles of topological rainbow trapping [49, 51] to perform robust signal processing have been explored only briefly so far [173]. Further, the signal processing implications of recent breakthroughs such as the development of "fractal" rainbow trapping (by graded quasicrystalline arrays [62]) have not yet been explored.

A final objective for the cochlea-inspired graded metamaterial community is to seek new, as yet unidentified stakeholders, in order to identify new opportunities and more diverse applications. For example, there are clear applications of these systems in solutions to hearing impairment (such as in hearing aids and cochlear implants). There could also be exciting applications in soundscape engineering and directional microphones.

Bibliography

[1] M. Abramowitz, I. A. Stegun, and R. H. Romer. *Handbook of Mathematical Functions with Formulas, Graphs, and Mathematical Tables*, volume 55 of *National Bureau of Standards Applied Mathematics Series*. U.S. Department of Commerce, 1948.

[2] D. Achlioptas. Database-friendly random projections: Johnson-Lindenstrauss with binary coins. *J. Comput. Syst. Sci.*, 66(4):671–687, 2003.

[3] H. Ammari, G. Ciraolo, H. Kang, H. Lee, and K. Yun. Spectral analysis of the Neumann–Poincaré operator and characterization of the stress concentration in anti-plane elasticity. *Arch. Ration. Mech. Anal.*, 208(1):275–304, 2013.

[4] H. Ammari and B. Davies. A fully coupled subwavelength resonance approach to filtering auditory signals. *Proc. R. Soc. A*, 475(2228):20190049, 2019.

[5] H. Ammari and B. Davies. Mimicking the active cochlea with a fluid-coupled array of subwavelength Hopf resonators. *Proc. R. Soc. A*, 476(2234):20190870, 2020.

[6] H. Ammari and B. Davies. Asymptotic links between signal processing, acoustic metamaterials, and biology. *SIAM J. Imaging Sci.*, 16(1):64–88, 2023.

[7] H. Ammari, B. Davies, and E. O. Hiltunen. Functional analytic methods for discrete approximations of subwavelength resonator systems. *arXiv preprint arXiv:2106.12301*, 2021.

[8] H. Ammari, B. Davies, and E. O. Hiltunen. Robust edge modes in dislocated systems of subwavelength resonators. *J. Lond. Math. Soc.*, 106(3):2075–2135, 2022.

[9] H. Ammari, B. Davies, E. O. Hiltunen, H. Lee, and S. Yu. High-order exceptional points and enhanced sensing in subwavelength resonator arrays. *Stud. Appl. Math.*, 146(2):440–462, 2021.

[10] H. Ammari, B. Davies, E. O. Hiltunen, H. Lee, and S. Yu. Exceptional points in parity–time-symmetric subwavelength metamaterials. *SIAM J. Math. Anal.*, 54(6):6223–6253, 2022.

[11] H. Ammari, B. Davies, E. O. Hiltunen, H. Lee, and S. Yu. Wave interaction with subwavelength resonators. In M. Chiappini and V. Vespri, editors, *Applied Mathematical Problems in Geophysics*, volume 2308 of *Lecture Notes in Mathematics, C.I.M.E. Foundation Subseries*, pages 23–83. Springer, 2022.

[12] H. Ammari, B. Davies, E. O. Hiltunen, and S. Yu. Topologically protected edge modes in one-dimensional chains of subwavelength resonators. *J. Math. Pures Appl.*, 144:17–49, 2020.

[13] H. Ammari, B. Davies, and S. Yu. Close-to-touching acoustic subwavelength resonators: eigenfrequency separation and gradient blow-up. *Multiscale Model. Simul.*, 18(3):1299–1317, 2020.

[14] H. Ammari, B. Fitzpatrick, H. Kang, M. Ruiz, S. Yu, and H. Zhang. *Mathematical and Computational Methods in Photonics and Phononics*, volume 235 of *Mathematical Surveys and Monographs*. American Mathematical Society, Providence, 2018.

[15] H. Ammari, B. Fitzpatrick, H. Lee, S. Yu, and H. Zhang. Subwavelength phononic bandgap opening in bubbly media. *J. Differ. Equ.*, 263(9):5610–5629, 2017.

[16] H. Ammari, B. Fitzpatrick, H. Lee, S. Yu, and H. Zhang. Double-negative acoustic metamaterials. *Q. Appl. Math.*, 77(4):767–791, 2019.

[17] H. Ammari and H. Kang. *Polarization and Moment Tensors: with Applications to Inverse Problems and Effective Medium Theory*, volume 162 of *Applied Mathematical Sciences*. Springer, 2007.

[18] H. Ammari, H. Kang, and H. Lee. *Layer Potential Techniques in Spectral Analysis*, volume 153 of *Mathematical Surveys and Monographs*. American Mathematical Society, Providence, 2009.

[19] H. Ammari, B. Li, and J. Zou. Mathematical analysis of electromagnetic scattering by dielectric nanoparticles with high refractive indices. *Trans. Am. Math. Soc.*, 2022.

[20] H. Ammari, M. Ruiz, S. Yu, and H. Zhang. Mathematical analysis of plasmonic resonances for nanoparticles: the full Maxwell equations. *J. Differ. Equ.*, 261(6):3615–3669, 2016.

[21] H. Ammari and H. Zhang. A mathematical theory of super-resolution by using a system of sub-wavelength Helmholtz resonators. *Commun. Math. Phys.*, 337(1):379–428, 2015.

https://doi.org/10.1515/9783110784961-009

[22] H. Ammari and H. Zhang. Effective medium theory for acoustic waves in bubbly fluids near Minnaert resonant frequency. *SIAM J. Math. Anal.*, 49(4):3252–3276, 2017.
[23] J. Andén and S. Mallat. Deep scattering spectrum. *IEEE Trans. Signal Process.*, 62(16):4114–4128, 2014.
[24] J. Andén, L. Sifre, S. Mallat, M. Kapoko, V. Lostanlen, and E. Oyallon. ScatNet v0 2. ScatNet v0.2. https://www.di.ens.fr/data/software/scatnet/download/, 2013. Accessed on 2021-10-21.
[25] T. Antonakakis, R. V. Craster, S. Guenneau, and E. A. Skelton. An asymptotic theory for waves guided by diffraction gratings or along microstructured surfaces. *Proc. R. Soc. A*, 470(2161):20130467, 2014.
[26] A. Arreola-Lucas, G. Baez, F. Cervera, A. Climente, R. A. Méndez-Sánchez, and J. Sánchez-Dehesa. Experimental evidence of rainbow trapping and Bloch oscillations of torsional waves in chirped metallic beams. *Sci. Rep.*, 9(1):1–13, 2019.
[27] R. Arriaga, D. Rutter, M. Cakmak, and S. Vempala. Visual categorization with random projection. *Neural Comput.*, 27(10):1–16, 2015.
[28] R. I. Arriaga and S. Vempala. An algorithmic theory of learning: Robust concepts and random projection. *Mach. Learn.*, 63:161–182, 2006.
[29] M. F. Ashby. Hybrids to fill holes in material property space. *Philos. Mag.*, 85(26–27):3235–3257, 2005.
[30] H. Attias and C. E. Schreiner. Temporal low-order statistics of natural sounds. In M. C. Mozer, M. Jordan, and T. Petsche, editors, *Advances in Neural Information Processing Systems 9*, pages 27–33. MIT Press, Cambridge, MA, 1997.
[31] H. Attias and C. E. Schreiner. Coding of naturalistic stimuli by auditory midbrain neurons. In M. C. Mozer, M. Jordan, M. Kearns, and S. Solla, editors, *Advances in Neural Information Processing Systems 10*, pages 103–109. MIT Press, Cambridge, MA, 1998.
[32] C. F. Babbs. Quantitative reappraisal of the Helmholtz-Guyton resonance theory of frequency tuning in the cochlea. *J. Biophys.*, 2011:1–16, 2011.
[33] M. Baraclough, I. R. Hooper, and W. L. Barnes. Investigation of the coupling between tunable split-ring resonators. *Phys. Rev. B*, 98(8):085146, 2018.
[34] A. Bell. A resonance approach to cochlear mechanics. *PLoS ONE*, 7(11):e47918, 2012.
[35] A. Bell, B. Davies, and H. Ammari. Bernhard Riemann, the ear, and an atom of consciousness. *Found. Sci.*, 27:855–873, 2022.
[36] A. Bell and H. P. Wit. The vibrating reed frequency meter: digital investigation of an early cochlear model. *PeerJ*, 3:e1333, 2015.
[37] A. Bell and H. P. Wit. Cochlear impulse responses resolved into sets of gammatones: the case for beating of closely spaced local resonances. *PeerJ*, 6:e6016, 2018.
[38] L. G. Bennetts, M. A. Peter, and R. V. Craster. Low-frequency wave-energy amplification in graded two-dimensional resonator arrays. *Philos. Trans. R. Soc. A*, 377(2156):20190104, 2019.
[39] A. Bensoussan, J. L. Lions, and G. Papanicolaou. *Asymptotic Analysis for Periodic Structures*. North-Holland, Amsterdam, 1978.
[40] J. M. Benyus. *Biomimicry: Innovation Inspired by Nature*. Morrow New York, 1997.
[41] A. D. Boardman, V. V. Grimalsky, Y. S. Kivshar, S. V. Koshevaya, M. Lapine, N. M. Litchinitser, V. N. Malnev, M. Noginov, Y. G. Rapoport, and V. M. Shalaev. Active and tunable metamaterials. *Laser Photonics Rev.*, 5(2):287–307, 2011.
[42] B. Boashash. Estimating and interpreting the instantaneous frequency of a signal. I. fundamentals. *Proc. IEEE*, 80(4):520–538, 1992.
[43] L. Brillouin. *Wave Propagation in Periodic Structures*. McGraw-Hill, 1946.
[44] N. D. Bruhin and B. Davies. Bioinspired random projections for robust, sparse classification. *SIAM J. Imaging Sci.*, 15(4):1833–1850, 2022.
[45] J. Bruna and S. Mallat. Invariant scattering convolution networks. *IEEE Trans. Pattern Anal. Mach. Intell.*, 35(8):1872–1886, 2013.
[46] S. Camalet, T. Duke, F. Jülicher, and J. Prost. Auditory sensitivity provided by self-tuned critical oscillations of hair cells. *Proc. Natl. Acad. Sci. USA*, 97(7):3183–3188, 2000.

[47] R. A. A. Campbell, H. Qin, K. Honegger, and W. Li. Imaging a population code for odor identity in the drosophila mushroom body. *J. Neurosci.*, 33(25):10568–10581, 2013.

[48] Centres for Disease Control and Prevention, U.S. Department of Health & Human Services. How does loud noise cause hearing loss? https://www.cdc.gov/nceh/hearing_loss/how_does_loud_noise_cause_hearing_loss.html, 2020. Accessed: 24-09-2021.

[49] G. J. Chaplain, J. M. De Ponti, G. Aguzzi, A. Colombi, and R. V. Craster. Topological rainbow trapping for elastic energy harvesting in graded Su-Schrieffer-Heeger systems. *Phys. Rev. Appl.*, 14(5):054035, 2020.

[50] G. J. Chaplain, A. S. Gliozzi, B. Davies, D. Urban, E. Descrovi, F. Bosia, and R. V. Craster. Tunable topological edge modes in Su–Schrieffer–Heeger arrays. *Appl. Phys. Lett.*, 122(22), 2023.

[51] G. J. Chaplain, D. Pajer, J. M. De Ponti, and R. V. Craster. Delineating rainbow reflection and trapping with applications for energy harvesting. *New J. Phys.*, 22(6):063024, 2020.

[52] A. Colombi, D. Colquitt, P. Roux, S. Guenneau, and R. V. Craster. A seismic metamaterial: The resonant metawedge. *Sci. Rep.*, 6(1):27717, 2016.

[53] R. V. Craster, T. Antonakakis, M. Makwana, and S. Guenneau. Dangers of using the edges of the Brillouin zone. *Phys. Rev. B*, 86(11):115130, 2012.

[54] R. V. Craster and B. Davies. Asymptotic characterisation of localised defect modes: Su–Schrieffer–Heeger and related models. *Multiscale Model. Simul.*, 21(3):827–848, 2023.

[55] R. V. Craster and S. Guenneau. *Acoustic Metamaterials: Negative Refraction, Imaging, Lensing and Cloaking*, volume 166 of *Springer Series in Materials Science*. Springer, London, 2013.

[56] R. V. Craster, J. Kaplunov, and A. V. Pichugin. High-frequency homogenization for periodic media. *Proc. R. Soc. A*, 466(2120):2341–2362, 2010.

[57] V. F. Dal Poggetto. Bioinspired acoustic metamaterials: From natural designs to optimized structures. *Front. Mater.*, 10:1176457, 2023.

[58] V. F. Dal Poggetto, F. Bosia, D. Urban, P. H. Beoletto, J. Torgersen, N. M. Pugno, and A. S. Gliozzi. Cochlea-inspired tonotopic resonators. *Mater. Des.*, 227:111712, 2023.

[59] P. Dallos. Cochlear neurobiology. In P. Dallos, A. N. Popper, and R. R. Fay, editors, *The Cochlea*, pages 1–43. Springer, New York, 1996.

[60] S. Dasgupta, C. F. Stevens, and S. Navlakha. A neural algorithm for a fundamental computing problem. *Science*, 358(6364):793–796, 2017.

[61] S. Davaria and P. A. Tarazaga. Toward developing arrays of active artificial hair cells. In *Special Topics in Structural Dynamics & Experimental Techniques, Volume 5: Proceedings of the 39th IMAC, A Conference and Exposition on Structural Dynamics 2021*, pages 75–80. Springer, 2022.

[62] B. Davies, G. J. Chaplain, T. A. Starkey, and R. V. Craster. Graded quasiperiodic metamaterials perform fractal rainbow trapping. *arXiv preprint arXiv:2305.10520*, 2023.

[63] B. Davies, L. Fehertoi-Nagy, and H. Putley. On the problem of comparing graded metamaterials. *Proc. R. Soc. A*, 479(2277):20230537, 2023.

[64] B. Davies and L. Herren. Robustness of subwavelength devices: a case study of cochlea-inspired rainbow sensors. *Proc. R. Soc. A*, 478(2262):20210765, 2022.

[65] J. M. De Ponti. *Graded Elastic Metamaterials for Energy Harvesting*. PoliM SpringerBriefs. Springer, 2021.

[66] J. M. De Ponti, A. Colombi, R. Ardito, F. Braghin, A. Corigliano, and R. V. Craster. Graded elastic metasurface for enhanced energy harvesting. *New J. Phys.*, 22(1):013013, 2020.

[67] M. Devaud, T. Hocquet, J. C. Bacri, and V. Leroy. The Minnaert bubble: an acoustic approach. *Eur. J. Phys.*, 29(6):1263, 2008.

[68] R. A. Diaz and W. J. Herrera. The positivity and other properties of the matrix of capacitance: Physical and mathematical implications. *J. Electrost.*, 69(6):587–595, 2011.

[69] G. S. Donaldson and R. A. Ruth. Derived band auditory brain-stem response estimates of traveling wave velocity in humans. i: Normal-hearing subjects. *J. Acoust. Soc. Am.*, 93(2):940–951, 1993.

[70] D. L. Donoho. Compressed sensing. *IEEE Trans. Inf. Theory*, 52(4):1289–1306, 2006.

[71] T. Duke and F. Jülicher. Active traveling wave in the cochlea. *Phys. Rev. Lett.*, 90(15):158101, 2003.

[72] T. Duke and F. Jülicher. Critical oscillators as active elements in hearing. In G. A. Manley, R. R. Fay, and A. N. Popper, editors, *Active Processes and Otoacoustic Emissions in Hearing*, pages 63–92. Springer, New York, 2008.

[73] H.-S. Ee and R. Agarwal. Tunable metasurface and flat optical zoom lens on a stretchable substrate. *Nano Lett.*, 16(4):2818–2823, 2016.

[74] V. M. Eguíluz, M. Ospeck, Y. Choe, A. J. Hudspeth, and M. O. Magnasco. Essential nonlinearities in hearing. *Phys. Rev. Lett.*, 84(22):5232, 2000.

[75] C. L. Fefferman, J. P. Lee-Thorp, and M. I. Weinstein. Topologically protected states in one-dimensional continuous systems and Dirac points. *Proc. Natl. Acad. Sci. USA*, 111(24):8759–8763, 2014.

[76] B. U. Felderhof and R. B. Jones. Addition theorems for spherical wave solutions of the vector Helmholtz equation. *J. Math. Phys.*, 28(4):836–839, 1987.

[77] F. Feppon and H. Ammari. Homogenization of sound-soft and high-contrast acoustic metamaterials in subcritical regimes. *ESAIM: Mathematical Model. Num.*, 57(2):491–543, 2023.

[78] J. L. Flanagan. Parametric coding of speech spectra. *J. Acoust. Soc. Am.*, 68(2):412–419, 1980.

[79] N. H. Fletcher. *Acoustic Systems in Biology*. Oxford University Press, New York, 1992.

[80] V. Galstyan, O. S. Pak, and H. A. Stone. A note on the breathing mode of an elastic sphere in Newtonian and complex fluids. *Phys. Fluids*, 27(3):032001, 2015.

[81] I. Giorgio, M. Spagnuolo, U. Andreaus, D. Scerrato, and A. M. Bersani. In-depth gaze at the astonishing mechanical behavior of bone: A review for designing bio-inspired hierarchical metamaterials. *Math. Mech. Solids*, 26(7):1074–1103, 2021.

[82] N. Goel, G. Bebis, and A. Nefian. Face recognition experiments with random projection. *Proc. SPIE*, 5779, Biometric Technology for Human Identification II, 2005.

[83] I. Gohberg and J. Leiterer. *Holomorphic Operator Functions of One Variable and Applications: Methods from Complex Analysis in Several Variables*, volume 192 of *Operator Theory Advances and Applications*. Birkhäuser, Basel, 2009.

[84] T. Gold. Hearing. II. the physical basis of the action of the cochlea. *P. Roy. Soc. Lond. B Bio.*, 135(881):492–498, 1948.

[85] G. H. Golub and C. F. Van Loan. *Matrix Computations*. Johns Hopkins University Press, Baltimore, 3rd edition, 1983.

[86] H. L. F. von Helmholtz. *On the Sensations of Tone as a Physiological Basis for the Theory of Music*. Longmans, Green, London, 1875.

[87] M. J. Hewitt and R. Meddis. A computer model of amplitude-modulation sensitivity of single units in the inferior colliculus. *J. Acoust. Soc. Am.*, 95(4):2145–2159, 1994.

[88] A. J. Hudspeth. The hair cells of the inner ear. *Sci. Am.*, 248(1):54–65, 1983.

[89] A. J. Hudspeth. Making an effort to listen: mechanical amplification in the ear. *Neuron*, 59(4):530–545, 2008.

[90] A. J. Hudspeth, F. Jülicher, and P. Martin. A critique of the critical cochlea: Hopf—a bifurcation—is better than none. *J. Neurophysiol.*, 104(3):1219–1229, 2010.

[91] S.-G. Hwang. Cauchy's interlace theorem for eigenvalues of Hermitian matrices. *Am. Math. Mon.*, 111(2):157–159, 2004.

[92] The Biomimicry Institute. Ask Nature. https://asknature.org [Online; accessed 12-March-2023].

[93] D. W. Jordan and P. Smith. *Nonlinear Ordinary Differential Equations: an Introduction to Dynamical Systems*, volume 2. Oxford University Press, Oxford, 1999.

[94] B. S. Joyce and P. A. Tarazaga. Mimicking the cochlear amplifier in a cantilever beam using nonlinear velocity feedback control. *Smart Mater. Struct.*, 23(7):075019, 2014.

[95] B. S. Joyce and P. A. Tarazaga. Developing an active artificial hair cell using nonlinear feedback control. *Smart Mater. Struct.*, 24(9):094004, 2015.

[96] B. S. Joyce and P. A. Tarazaga. A study of active artificial hair cell models inspired by outer hair cell somatic motility. *J. Intell. Mater. Syst. Struct.*, 28(6):811–823, 2017.

[97] F. Jülicher, D. Andor, and T. Duke. Physical basis of two-tone interference in hearing. *Proc. Natl. Acad. Sci. USA*, 98(16):9080–9085, 2001.

[98] M. Kadic, G. W. Milton, M. van Hecke, and M. Wegener. 3d metamaterials. *Nat. Rev. Phys.*, 1(3):198–210, 2019.

[99] I. V. Kamotski and V. P. Smyshlyaev. Two-scale homogenization for a general class of high contrast PDE systems with periodic coefficients. *Appl. Anal.*, 98(1–2):64–90, 2019.

[100] K. Kanders and R. Stoop. Spontaneous otoacoustic emissions from higher order signal coupling. In L. Fortuna, A. Buscarino, and R. Stoop, editors, *Advances on Nonlinear Dynamics of Electronic Systems*, volume 17, pages 103–108. World Scientific, 2019.

[101] A. Karlos and S. J. Elliott. Cochlea-inspired design of an acoustic rainbow sensor with a smoothly varying frequency response. *Sci. Rep.*, 10(1):1–11, 2020.

[102] D. T. Kemp. Otoacoustic emissions, their origin in cochlear function, and use. *Br. Med. Bull.*, 63(1):223–241, 2002.

[103] A. Kern and R. Stoop. Essential role of couplings between hearing nonlinearities. *Phys. Rev. Lett.*, 91(12):128101, 2003.

[104] A. B. Khanikaev, S. Hossein Mousavi, W.-K. Tse, M. Kargarian, A. H. MacDonald, and G. Shvets. Photonic topological insulators. *Nat. Mater.*, 12(3):233–239, 2013.

[105] P. Kuchment. *Floquet Theory for Partial Differential Equations*, volume 60 of *Operator Theory: Advances and Applications*. Birkhäuser Verlag, Basel, 1993.

[106] P. Kuchment. An overview of periodic elliptic operators. *Bull. Am. Math. Soc.*, 53(3):343–414, 2016.

[107] A. I. Kuznetsov, A. E. Miroshnichenko, M. L. Brongersma, Y. S. Kivshar, and B. Luk'yanchuk. Optically resonant dielectric nanostructures. *Science*, 354(6314):aag2472, 2016.

[108] V. Leroy, A. Bretagne, M. Fink, H. Willaime, P. Tabeling, and A. Tourin. Design and characterization of bubble phononic crystals. *Appl. Phys. Lett.*, 95(17):171904, 2009.

[109] V. Leroy, A. Strybulevych, M. G. Scanlon, and J. H. Page. Transmission of ultrasound through a single layer of bubbles. *Eur. Phys. J. E*, 29(1):123–130, 2009.

[110] K. D. Lerud, J. C. Kim, F. V. Almonte, L. H. Carney, and E. W. Large. A canonical oscillator model of cochlear dynamics. *Hear. Res.*, 380:100–107, 2019.

[111] N. A. Lesica and B. Grothe. Efficient temporal processing of naturalistic sounds. *PLoS ONE*, 3(2):e1655, 2008.

[112] C. M. Linton. Lattice sums for the Helmholtz equation. *SIAM Rev.*, 52(4):630–674, 2010.

[113] C. M. Linton and I. Thompson. One- and two-dimensional lattice sums for the three-dimensional Helmholtz equation. *J. Comput. Phys.*, 228(6):1815–1829, 2009.

[114] S. Lu, D. Mountain, and A. Hubbard. Is stereocilia velocity or displacement feedback used in the cochlear amplifier? In N. P. Cooper and D. T. Kemp, editors, *Concepts And Challenges In The Biophysics Of Hearing*, pages 297–302. World Scientific, 2009.

[115] R. F. Lyon, A. G. Katsiamis, and E. M. Drakakis. History and future of auditory filter models. In *Proceedings of 2010 IEEE International Symposium on Circuits and Systems*, pages 3809–3812. IEEE, 2010.

[116] R. F. Lyon. *Human and Machine Hearing*. Cambridge University Press, 2017.

[117] M. O. Magnasco. A wave traveling over a Hopf instability shapes the cochlear tuning curve. *Phys. Rev. Lett.*, 90(5):058101, 2003.

[118] S. Mallat. Group invariant scattering. *Commun. Pure Appl. Math.*, 65(10):1331–1398, 2012.

[119] S. Mallat. Understanding deep convolutional networks. *Philos. Trans. R. Soc. A*, 374(2065):20150203, 2016.

[120] P. Martin and A. J. Hudspeth. Compressive nonlinearity in the hair bundle's active response to mechanical stimulation. *Proc. Natl. Acad. Sci. USA*, 98(25):14386–14391, 2001.

[121] N. Y. Masse, G. C. Turner, and G. S. X. E. Jefferis. Olfactory information processing in drosophila. *Curr. Biol.*, 19(16):R700–R713, 2009.

[122] J. C. Maxwell. *A Treatise on Electricity and Magnetism, volume 1*. Clarendon Press, Oxford, 1873.

[123] J. H. McDermott and E. P. Simoncelli. Sound texture perception via statistics of the auditory periphery: evidence from sound synthesis. *Neuron*, 71(5):926–940, 2011.

[124] G. W. Milton and N. A. Nicorovici. On the cloaking effects associated with anomalous localized resonance. *Proc. R. Soc. A*, 462(2074):3027–3059, 2006.

[125] M. Miniaci, A. Krushynska, A. S. Gliozzi, N. Kherraz, F. Bosia, and N. M. Pugno. Design and fabrication of bioinspired hierarchical dissipative elastic metamaterials. *Phys. Rev. Appl.*, 10(2):024012, 2018.

[126] M. Miniaci, A. Krushynska, A. B. Movchan, F. Bosia, and N. M. Pugno. Spider web-inspired acoustic metamaterials. *Appl. Phys. Lett.*, 109(7):071905, 2016.

[127] M. Minnaert. On musical air-bubbles and the sounds of running water. *Philos. Mag.*, 16(104):235–248, 1933.

[128] A. H. Nayfeh and D. T. Mook. *Nonlinear Oscillations*. Wiley, New York, 1979.

[129] J.-C. Nédélec. New trends in the use and analysis of integral equations. In *Mathematics of Computation 1943–1993: a half-century of computational mathematics (Vancouver, BC, 1993)*, volume 48 of *Proc. Sympos. Appl. Math.*, pages 151–176. Amer. Math. Soc., Providence, RI, 1994.

[130] S. T. Neely and D. O. Kim. A model for active elements in cochlear biomechanics. *J. Acoust. Soc. Am.*, 79(5):1472–1480, 1986.

[131] T. R. Neil, Z. Shen, D. Robert, B. W. Drinkwater, and M. W. Holderied. Moth wings are acoustic metamaterials. *Proc. Natl. Acad. Sci. USA*, 117(49):31134–31141, 2020.

[132] I. Nelken, Y. Rotman, and O. B. Yosef. Responses of auditory-cortex neurons to structural features of natural sounds. *Nature*, 397(6715):154–157, 1999.

[133] E. S. Olson. Direct measurement of intra-cochlear pressure waves. *Nature*, 402(6761):526, 1999.

[134] C. H. Papadimitriou and S. S. Vempala. Random projection in the brain and computation with assemblies of neurons. In A. Blum, editor, *10th Innovations in Theoretical Computer Science Conference (ITCS 2019)*, volume 124 of *Leibniz International Proceedings in Informatics (LIPIcs)*, pages 57:1–57:19, Dagstuhl, Germany, 2018. Schloss Dagstuhl–Leibniz-Zentrum fuer Informatik.

[135] W. Park and J.-B. Lee. Mechanically tunable photonic crystal structure. *Appl. Phys. Lett.*, 85(21):4845–4847, 2004.

[136] R. D. Patterson, I. Nimmo-Smith, J. Holdsworth, and P. Rice. *APU report 2341: An efficient auditory filterbank based on the gammatone function*. Applied Psychology Unit, Cambridge, 1988.

[137] J. B. Pendry. Negative refraction makes a perfect lens. *Phys. Rev. Lett.*, 85(18):3966, 2000.

[138] I. M. Pryce, K. Aydin, Y. A. Kelaita, R. M. Briggs, and H. A. Atwater. Highly strained compliant optical metamaterials with large frequency tunability. *Nano Lett.*, 10(10):4222–4227, 2010.

[139] A. Rahimi and B. Recht. Random features for large-scale kernel machines. In J. Platt, D. Koller, Y. Singer, and S. Roweis, editors, *Advances in Neural Information Processing Systems*, volume 20. Curran Associates, Inc., 2007.

[140] A. Rahimi and B. Recht. Weighted sums of random kitchen sinks: Replacing minimization with randomization in learning. In D. Koller, D. Schuurmans, Y. Bengio, and L. Bottou, editors, *Advances in Neural Information Processing Systems*, volume 21. Curran Associates, Inc., 2008.

[141] R. H. Rand. Lecture notes on nonlinear vibrations, version 53, 2012. URL: https://hdl.handle.net/1813/28989.

[142] M. C. Rechtsman, J. M. Zeuner, Y. Plotnik, Y. Lumer, D. Podolsky, F. Dreisow, S. Nolte, M. Segev, and A. Szameit. Photonic Floquet topological insulators. *Nature*, 496(7444):196–200, 2013.

[143] T. Reichenbach and A. J. Hudspeth. The physics of hearing: fluid mechanics and the active process of the inner ear. *Rep. Prog. Phys.*, 77(7):076601, 2014.

[144] L. Robles, M. A. Ruggero, and N. C. Rich. Two-tone distortion on the basilar membrane of the chinchilla cochlea. *J. Neurophysiol.*, 77(5):2385–2399, 1997.

[145] M. A. Ruggero, L. Robles, and N. C. Rich. Two-tone suppression in the basilar membrane of the cochlea: Mechanical basis of auditory-nerve rate suppression. *J. Neurophysiol.*, 68:1087–1099, 1992.

[146] M. Rupin, G. Lerosey, J. de Rosny, and F. Lemoult. Mimicking the cochlea with an active acoustic metamaterial. *New J. Phys.*, 21:093012, 2019.

[147] O. Schnitzer. Waves in slowly varying band-gap media. *SIAM J. Appl. Math.*, 77(4):1516–1535, 2017.
[148] E. A. Skelton, R. V. Craster, A. Colombi, and D. J. Colquitt. The multi-physics metawedge: graded arrays on fluid-loaded elastic plates and the mechanical analogues of rainbow trapping and mode conversion. *New J. Phys.*, 20(5):053017, 2018.
[149] C. F. Stevens. What the fly's nose tells the fly's brain. *Proc. Natl. Acad. Sci. USA*, 112(30):9460–9465, 2015.
[150] C. F. Stevens. A statistical property of fly odor responses is conserved across odors. *Proc. Natl. Acad. Sci. USA*, 113(24):6737–6742, 2016.
[151] J. J. Stoker. *NoNlinear Vibrations in Mechanical and Electrical Systems*, volume 2. Interscience Publishers, New York, 1950.
[152] S. H. Strogatz. *NOnlinear Dynamics and Chaos: With Applications to Physics, Biology, Chemistry, and Engineering*. Perseus, Reading, MA, 1994.
[153] W. P. Su, J. R. Schrieffer, and A. J. Heeger. Solitons in polyacetylene. *Phys. Rev. Lett.*, 42:1698–1701, Jun 1979.
[154] H. Tao, A. C. Strikwerda, K. Fan, W. J. Padilla, X. Zhang, and R. D. Averitt. Reconfigurable terahertz metamaterials. *Phys. Rev. Lett.*, 103(14):147401, 2009.
[155] T. Tao. *Topics in Random Matrix Theory*, volume 132 of *Graduate Studies in Mathematics*. American Mathematical Society, 2012.
[156] T. Tao and V. Vu. On random ± 1 matrices: singularity and determinant. *Random Struct. Algorithms*, 28(1):1–23, 2006.
[157] F. E. Theunissen and J. E. Elie. Neural processing of natural sounds. *Nat. Rev. Neurosci.*, 15(6):355–366, 2014.
[158] K. L. Tsakmakidis, A. D. Boardman, and O. Hess. 'Trapped rainbow' storage of light in metamaterials. *Nature*, 450:397–401, 2007.
[159] G. Tzanetakis and P. Cook. Musical genre classification of audio signals. *IEEE Trans. Speech Audio Process.*, 10(5):293–302, 2002.
[160] V. G. Veselago. The electrodynamics of substances with simultaneously negative values of ϵ and μ. *Sov. Phys. Usp.*, 10(4):509–514, 1968.
[161] G. von Békésy. *Experiments in Hearing*. McGraw-Hill, New York, 1960.
[162] R. F. Voss and J. Clarke. '1/f noise' in music and speech. *Nature*, 258(5533):317–318, 1975.
[163] D. Wagg and S. A. Neild. *Nonlinear Vibration with Control*. Springer, Cham, 2016.
[164] S. Walia, C. M. Shah, P. Gutruf, H. Nili, D. R. Chowdhury, W. Withayachumnankul, M. Bhaskaran, and S. Sriram. Flexible metasurfaces and metamaterials: A review of materials and fabrication processes at micro-and nano-scales. *Appl. Phys. Rev.*, 2(1):011303, 2015.
[165] R. M. Walser. Metamaterials: What are they? What are they good for? In *APS March Meeting Abstracts*, APS Meeting Abstracts, page Z5.001, March 2000.
[166] E. P. Wigner. *On the Matrices Which Reduce the Kronecker Products of Representations of SR Groups*. Springer, 1993.
[167] Wikipedia contributors. Metamaterial — Wikipedia, the free encyclopedia, 2022. https://en.wikipedia.org/wiki/Metamaterial [Online; accessed 12-October-2022].
[168] S. M. N. Woolley, T. E. Fremouw, A. Hsu, and F. E. Theunissen. Tuning for spectro-temporal modulations as a mechanism for auditory discrimination of natural sounds. *Nat. Neurosci.*, 8(10):1371–1379, 2005.
[169] P.-Z. Wu, J. T. O'Malley, V. de Gruttola, and M. C. Liberman. Age-related hearing loss is dominated by damage to inner ear sensory cells, not the cellular battery that powers them. *J. Neurosci.*, 40(33):6357–6366, 2020.
[170] S. Xiao, T. Wang, T. Liu, C. Zhou, X. Jiang, and J. Zhang. Active metamaterials and metadevices: a review. *J. Phys. D, Appl. Phys.*, 53(50):503002, 2020.
[171] Z. Yang, F. Gao, X. Shi, X. Lin, Z. Gao, Y. Chong, and B. Zhang. Topological acoustics. *Phys. Rev. Lett.*, 114(11):114301, 2015.

[172] J. Zak. Berry's phase for energy bands in solids. *Phys. Rev. Lett.*, 62(23):2747, 1989.

[173] F. Zangeneh-Nejad and R. Fleury. Topological analog signal processing. *Nat. Commun.*, 10(1):2058, 2019.

[174] B. Zhao, H. R. Thomsen, J. M. De Ponti, E. Riva, B. Van Damme, A. Bergamini, E. Chatzi, and A. Colombi. A graded metamaterial for broadband and high-capability piezoelectric energy harvesting. *Energy Convers. Manag.*, 269:116056, 2022.

[175] L. Zhao and S. Zhou. Compact acoustic rainbow trapping in a bioinspired spiral array of graded locally resonant metamaterials. *Sensors*, 19(4):788, 2019.

[176] J. Zhu, Y. Chen, X. Zhu, F. J. Garcia-Vidal, X. Yin, W. Zhang, and X. Zhang. Acoustic rainbow trapping. *Sci. Rep.*, 3:1728, 2013.

[177] P. M. Zurek. Acoustic emissions from the ear: A summary of results from humans and animals. *J. Acoust. Soc. Am.*, 78(1):340–344, 1985.

Index

https://doi.org/10.1515/9783110784961-010

De Gruyter Series in Applied and Numerical Mathematics

Volume 8
Zhi-Zhong Sun, Qifeng Zhang, Guang-hua Gao
Finite Difference Methods for Nonlinear Evolution Equations, 2023
ISBN 978-3-11-079585-1, e-ISBN (PDF) 978-3-11-079601-8, e-ISBN (EPUB) 978-3-11-079611-7

Volume 7
Roland Glowinski, Tsorng-Whay Pan
Numerical Simulation of Incompressible Viscous Flow. Methods and Applications, 2022
ISBN 978-3-11-078491-6, e-ISBN (PDF) 978-3-11-078501-2, e-ISBN (EPUB) 978-3-11-078505-0

Volume 6
Bruno Després
Neural Networks and Numerical Analysis, 2022
ISBN 978-3-11-078312-4, e-ISBN (PDF) 978-3-11-078318-6, e-ISBN (EPUB) 978-3-11-078326-1

Volume 5/1
Alexander V. Bobylev
Kinetic Equations. Volume 1: Boltzmann Equation, Maxwell Models, and Hydrodynamics beyond Navier–Stokes, 2020
ISBN 978-3-11-055012-2, e-ISBN (PDF) 978-3-11-055098-6, e-ISBN (EPUB) 978-3-11-055017-7

Volume 4
Claude Le Bris, Pierre-Louis Lions
Parabolic Equations with Irregular Data and Related Issues. Applications to Stochastic Differential Equations, 2019
ISBN 978-3-11-063313-9, e-ISBN (PDF) 978-3-11-063550-8, e-ISBN (EPUB) 978-3-11-063314-6

Volume 3
Dominic Breit, Eduard Feireisl, Martina Hofmanová
Stochastically Forced Compressible Fluid Flows, 2018
ISBN 978-3-11-049050-3, e-ISBN (PDF) 978-3-11-049255-2, e-ISBN (EPUB) 978-3-11-049076-3

Volume 2
Zahari Zlatev, Ivan Dimov, István Faragó, Ágnes Havasi
Richardson Extrapolation. Practical Aspects and Applications, 2017
ISBN 978-3-11-051649-4, e-ISBN (PDF) 978-3-11-053300-2, e-ISBN (EPUB) 978-3-11-053198-5

Volume 1
Anvarbek Meirmanov, Oleg V. Galtsev, Reshat N. Zimin
Free Boundaries in Rock Mechanics, 2017
ISBN 978-3-11-054490-9, e-ISBN (PDF) 978-3-11-054616-3, e-ISBN (EPUB) 978-3-11-054504-3

www.degruyter.com

www.ingramcontent.com/pod-product-compliance
Ingram Content Group UK Ltd.
Pitfield, Milton Keynes, MK11 3LW, UK
UKHW052008051026
15624UKWH00004B/631

9 783110 784046